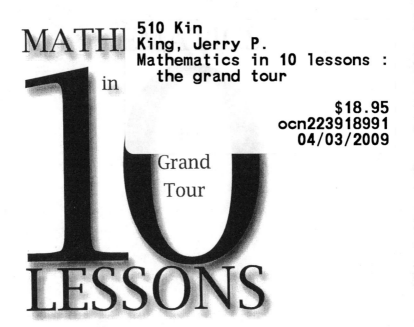

MATH in

Grand
Tour

10 LESSONS

D1369978

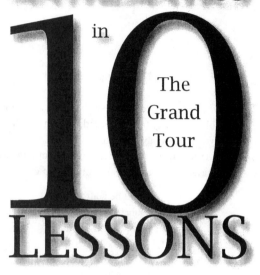

MATHEMATICS in 10 LESSONS

The Grand Tour

JERRY P. KING

Prometheus Books

59 John Glenn Drive
Amherst, New York 14228-2119

Published 2009 by Prometheus Books

Inquiries should be addressed to
Prometheus Books
59 John Glenn Drive
Amherst, New York 14228–2119
VOICE: 716–691–0133, ext. 210
FAX: 716–691–0137
WWW.PROMETHEUSBOOKS.COM

13 12 11 10 09   5 4 3 2 1

Library of Congress Cataloging-in-Publication Data

King, Jerry P.
    Mathematics in 10 lessons : the grand tour / by Jerry P. King.
        p. cm.
    Includes bibliographical references and index.
    ISBN 978–1–59102–686–0 (pbk. : alk. paper)
    1. Mathematics—Popular works. I. Title, II. Title: Mathematics in ten lessons.

QA93.K535 2009
510—dc22

2008049258

Printed in the United States on acid-free paper

CONTENTS

ACKNOWLEDGMENTS

Somewhere in this book I confess that I am no longer sure as to what I discovered for myself and what I have learned from others. Perhaps I never knew. Maybe no one knows. But I am sure that all the mathematics I wrote in this book is mathematics I learned from someone else. I created none of it. Standing behind me is a misty line of mathematicians and teachers of mathematics who separately showed me parts of this lovely subject and motivated me to learn some small piece of it. I owe them all a debt I can never repay. The line is not infinite but it is long and I cannot list their names. But they are not forgotten. I remember them all. If I am lucky one or two of them will remember me.

But there are names I must mention. Three of my colleagues: Wei-Min Huang, Clifford Queen, and Ramamirtham Venkataraman read portions of the manuscript and made helpful suggestions. I thank them all. Graduate student Jiahuan Kang provided expert, indispensable technical help. He has my thanks. At Prometheus Books, Production Manager Christine Kramer was extremely cooperative with regard to the submission of a complicated manuscript. I am greatly appreciative.

Then there is my editor, Linda Regan. Linda's extraordinary competence is well known as all who have worked with her will testify. But here she went beyond competence. She gave me suggestions, encouragement, and, sometimes, deep ideas. Most of all, she showed great patience. It is a flat truth that without Linda this book would not exist.

My indebtedness to Jane is not countable. I owe her everything.

HOW TO READ THIS BOOK

A poet once told me that a poem must be read twice. First you discover what the poem says. Then you learn its way of saying. Each poem, the poet said, consists of these two: *the thing said and the way of saying.*

The first reading is usually the easier. Here, you learn the poem's subject and what is said about it. The second time, you read the poem for its structure. You deal with the technical aspects of the poem: rhyme and rhythm, metaphor and syntax. When you become good at it, the two readings blend so that the thing said and the way of saying merge. And it is this merging that gives the poem integrity.

But a beginner needs to separate clearly the two readings. A beginner must often consciously struggle against a poem's tendency to hide its own subject behind layers of metaphor and simile. At first glance the subject may seem transparent. The poem may appear to be only a literal description of the morning flight of a falcon. But is it? Perhaps the transparency is illusionary. The sweeping falcon may portray nothing less than a deity coming down and making the mountains smoke. To find the true subject, you may have to peel away layers of analogies one by one, like the concentric shells of an onion. This may not be easy but it will be enriching. As it occurs, you are necessarily led into the second reading of the poem: the way of saying.

While the first reading is largely descriptive, the second is technical. Here you read slowly, repetitively, and often discontinuously. You may set the poem aside, let it settle, and go back to it several times until you master the technicalities and the subtleties of its way of saying. This mastering is required if the poem is to be truly read. Failure means that you have not discerned the poem's essence—you have failed to distinguish poetry from prose. You must read both descriptively and technically. Poetry can be read no other way. Neither can mathematics.

Mathematics—like poetry—divides into the thing said and the way of saying. Also, as in poetry, the beginner is well advised to distinguish between the two. In mathematics, *the thing said* ordinarily consists of declarative statements that take the form of definitions or theorems. These statements are set down in the subject's abstract and symbolic language. *The way of saying* in mathematics consists of the technical manipulation of these symbols so as to produce *proofs* of the theorems. Unlike poetry, the statements in mathematics have precise meanings. Words in the theorems have exactly the meanings given in the definitions. There are no analogies or metaphors to peel away. In this sense, mathematics is easier than poetry. On the other hand, the way of saying in mathematics also follows certain precisely described rules. In the proof of any particular theorem, mathematical symbols may be manipulated only according to rules that are described in the definitions or that have been established by previous theorems. Mathematics is precise or else it is nothing. The commitment to precision makes mathematics more difficult than poetry.

This book contains real mathematics. You will find precision in the form of definitions, theorems, and proofs. Also, you will find exposition, explanation, and deliberate imprecision. The idea is to embed the mathematics within sufficient description so as to make it palatable. Moreover, the book is almost self-contained. Essentially no prior mathematical knowledge is assumed. Thus a mathematical idea on any particular page can be understood through the ideas on the preceding pages. Inevitably you will come to an idea that you do not understand. Read on. The troublesome point may be explained in the next paragraph, or on the next page.

Read this book as you would poetry. First, read the exposition and the statements. Learn what the mathematics says. Then go at the proofs and master the way of saying. There is no need to rush the second reading. Put the book aside and let the ideas settle. The mathematics will not go away.

When the *Iliad* is forgotten and the Parthenon is dust, mathematics will remain. Nothing ever created by man is more permanent.

INTRODUCTION

Mathematicians know two things that others do not. They know first that all mathematics flows from a few *fundamental principles*. Second, they know that *aesthetic* considerations provide both the motivation for mathematics research and the standards for evaluating the research once it is done. The fundamental principles constitute the source of mathematics, and the pulling force of beauty makes it go.

Of course, mathematicians know more than this. But much of their additional knowledge is shared by others, particularly by scientists and engineers. These others—those who *use* mathematics—are well acquainted with the subject's *utility*. But many scientists and engineers know neither of the aesthetics of mathematics nor the existence of the principles on which the subject stands. Mathematicians know these two things. Only mathematicians know.

Engineers generally think mathematicians create mathematics—or ought to—because it is useful. They believe the subject consists of a thick set of complicated rules that must be learned the way a medical student learns the contents of *Gray's Anatomy*. Once learned—the engineers think—the rules allow you to manipulate algebraic symbols across blank sheets of paper so that something called mathematics is produced. To these engineers—and to almost everyone outside the discipline of pure mathematics—the rules are indistinguishable from the mathematics. Nonmathematicians are unaware that each complicated rule can be produced from simpler rules. And these, in turn, from laws even simpler; and on and on until one comes back to a handful of fundamental rules that are so simple their correctness cannot be doubted.

In this book the term *mathematician* means *a person who does mathematical research* and this means *a person who creates new mathematics*. Thus by *mathematician* I mean a person who creates new mathematics. Ordinarily a mathematician holds a professorial position in a research university and his academic success depends mainly on the quality and the quantity of the mathematics he produces. Most research mathematicians are also

11

teachers of mathematics. But not all mathematics teachers are research mathematicians.

A mathematician memorizes few rules. Instead he learns certain fundamental principles and from these principles he *derives* other rules as they are needed. When he is done and mathematics lies before him, set down on the page in the strange and beautiful language of symbols and curves, the mathematician knows from whence it came. No matter how complicated the final result, the mathematician can if he wishes trace it back through reverse evolution to a handful of self-evident axioms.

As for aesthetics and mathematics, the connection between them is essentially unknown outside the closed ring of practicing mathematicians. To most people the notion that mathematics possesses an aesthetic component is as remote as the dark side of the moon. Mathematicians, however, see mathematics as high art. Mathematicians do not *do mathematics* because it is useful. Mathematicians do mathematics because mathematics is lovely.

And it is precisely this knowledge of the aesthetic value of mathematics, and of the principles on which the subject stands, that defines the mathematician and separates him from the rest of intellectual society. Only this and nothing more.

In a book called *The Art of Mathematics*,[1] I presented, to an intended audience of those in the humanities, the notion of mathematics as art. There, I discussed the importance mathematicians place on aesthetic considerations as they create, and then evaluate, mathematics. "The motivation for mathematics is beauty," I wrote. "And so is its evaluation. The highest praise one mathematician gives to another's work is to pronounce it *elegant*."

The Art of Mathematics contains little mathematics. In that book I looked at mathematics from the outside as one looks at a Monet from a distance, focusing on the spectrum of colors rather than the brushstrokes. I was writing *about* mathematics rather than writing mathematics. I tried to describe—to a nontechnical audience—what it is that mathematicians do and why they do it. Consequently, I wrote description and exposition. I did not write mathematics.

The purpose of the present book is to present to a wider audience a volume containing a moderate amount of actual mathematics. I hope to reach an audience composed of two kinds of educated professionals: those who, for the first time, truly want to learn mathematics and those others who once possessed mathematical literacy but who—over the years—have seen their knowledge slip away. The presentation will be leisurely, nonthreatening, and essentially self-contained. And the topics are chosen so that the reader can grasp the mathematician's concept of *fundamental principles*.

Consequently, we begin our study at the most basic level. Almost no prior mathematical knowledge is assumed other than the reader's ability to point and count. Not even the ability to add simple fractions is presupposed.

Nor do I assume familiarity with the logical concepts by which mathematicians string together mathematical ideas. Our study, in fact, commences with an examination of the basic notions of symbolic logic. In particular, we study the fundamental concept of logical implication. Then we proceed to the counting numbers 1, 2, 3, . . . , and the surrounding area of mathematics called number theory. From number theory we move to mathematical analysis and the great collection of ideas known as calculus. Along the way we pause to examine the set of mathematical ideas called probability theory, which explains and predicts random phenomena such as the drift of electrons or the tossing of biased coins. We cover a fair amount of material but at each stage the purpose is to present basic ideas that illustrate the mathematician's concept of fundamental principles.

Obviously, we cannot—at this level—come to grips completely with the foundations of mathematics: on the one hand, they lie too deep within the subject; and, on the other, their considerations are simply too lofty. At either end, they fall beyond our reach. But we can—through properly chosen topics—illustrate the existence of fundamental mathematical principles and show how this allows the mathematician to go about the subject differently from the rest of us. The subject seems easier to a mathematician. Why?

Frequently I teach freshman calculus. I teach out of a textbook (chosen by someone else) the size of a cinderblock. The book weighs six pounds. Let it slip and it will break your foot. Students lug the book around and even read parts of it from time to time. I go to class without the book. I walk into the classroom armed only with a pristine stick of white chalk. Then I simultaneously lecture and write mathematics on the blackboard for fifty minutes, with no notes and almost without pause. This is no great feat. But the students—believing like most that mathematics consists only of a complicated set of rules—are often overly impressed with my memory. They believe I have memorized the textbook. They think I expect them to memorize it too.

They are wrong. Neither I nor any of my colleagues have a textbook in our heads. Instead, we have learned some principles—in this case, principles of calculus. From these laws the lecture flows as inexorably as the Nile flows to the sea. Memory plays a small role.

Take my clunky calculus text. Open it to an arbitrary page and point to a complicated formula on that page. (You can be sure to find one; it will be set out in multicolor and framed in a box.) Now go find a mathematician. Ask him to say the formula from his memory.

You will likely find that he cannot. He does not carry the result in his head. But you will also find that he can, in a reasonable time, *derive* the formula for you. He will produce it as you watch. He'll begin with something more basic, something he does carry in his head. And, as you watch him write, you will see the formula appear. It will show vaguely at first, and then gradually as he makes a false start or so. Then it will come into existence, symbol by symbol, exactly as in the book. When he's properly warmed up, the mathematics tumbles out as if it had a life of its own.

And that's just for starters: give the mathematician sufficient time and he'll reproduce the entire textbook. Not word for word, of course, but close enough. (The result might even be a better book.)

The lesson to be learned is this: the mathematician has mastered calculus by learning certain fundamentals very well. He has absorbed the basics of the subject exactly and from these basics he can produce the rest of it whenever he must. In fact, he has learned all his mathematics in precisely the same way.

So should we. And the present book is a step in that direction. The thesis is that *the way to learn mathematics is to learn some fundamental parts of it truly and well*. Once this has been done, any other part of mathematics that must be learned can be learned.

The book's topics and the manner of presentation have been chosen with this thesis in mind. The overriding theme is: *the precise learning of basic ideas provides a springboard for the learning of more complex ideas*. I am certain that this is the way mathematicians learn mathematics. And I am also certain that it is the easiest way to learn the subject. It may, in fact, be the only way.

Much of the power of mathematics comes from the symbols in which it is written. Each symbol expresses precisely defined mathematical ideas. A major part of learning mathematics consists in learning the symbols, the ideas they contain, and the rules for manipulating them to produce more powerful ideas. Those new to mathematics will often find the symbolism off-putting. But there is no avoiding it. The symbolism must be mastered if mathematics is to be mastered. As we proceed we will introduce many symbols. A key to symbols—found near the back of the book on page 385—has been produced to assist the reader. The key provides—in each case—the symbol itself, how it is read, the page on which it is precisely defined, and an intuitive description of what the symbol means. (As you use the key, keep in mind that it is the precise definition that counts.)

In his famous Rede Lecture at Cambridge in 1959, C. P. Snow asserted that Western intellectual society has become separated into two disparate groups. One of these groups, Snow claimed, contains the scientists and the other is composed of literary intellectuals. Snow called these groups "The Two Cultures." Lord Snow deplored the separation because he found that meaningful communication between the cultures does not exist. Even worse, Snow claimed, the highly educated people in one of the cultures simply do not like those in the other. Between the cultures, Snow said, lives "hostility and dislike."[2]

Undoubtedly, Lord Snow's assertion of the existence of the two cultures is correct. All a doubter need do is go to any university campus and look. The two cultures will find you before you find them. Ask a scientist directions to the Hall of Liberal Arts and he will likely end his set of directions with a gratuitous remark about the fuzzy-mindedness of the humanists who work there. When you arrive you'll find, outside the English Department door, bulletin board cartoons lampooning the illiteracy of professors who wear white laboratory coats and peer into microscopes. On campus, the separateness of the two cultures will jump out at you whether you are seeking it or not.

However, I do not believe that the cultures are defined by something intrinsic to the nature of humanists or to the nature of scientists. Rather, the cultures are determined precisely by the presence or absence of a certain level of mathematical proficiency. I have referred to the cultures as being composed, respectively, of people of Type M or of people of Type N—a Type M person being one who knows a certain amount of mathematics and a Type N person being one who does not.[3]

Thus, I claimed that the separating agent between the cultures is mathematics. The M-culture contains Snow's scientists; the N-culture contains his literary intellectuals. A physicist is type M; a poet is type N. The physicist knows mathematics, the poet does not. Because of this, they do not communicate.

The way to bring together a separating pair is to identify the cause of the separation and then to do something about it. As in the case of a married couple drifting apart, neither the identification nor the cure may be easy. However, there exists an entire industry of psychologists, counselors, and therapists dedicated to the investigation of causes and cures for separation and divorce. The people who pursue this work are ready and available. For a price they will help you with your marital problems.

But little help is available to the poet or the physicist. The separating agent between them is mathematics and the corresponding counseling industry should be a collection of practicing mathematicians. But it is not.

Mathematicians concentrate on research. Research is what they want to do and mainly what universities pay them to do. Teaching is often not a high priority and the culture distinction is not a priority at all. By and large, mathematicians consider the culture gap inevitable. Poets are poets and physicists are physicists. It is natural that there be no commerce between them.

I have argued that, although mathematics stands as the separating agent between the two cultures, it can be transformed into a cure. But, as in the case of any other crippling national disease, the curing process will not be easy. Nor will it be quick. What is required is nothing less than a revolution in the teaching of mathematics at all levels. What you must do is redesign the education system so that, early on, mathematics is presented to both potential poets and future scientists in such a manner that they each see the subject as the mathematicians do—as high art. This will require nothing less than a reeducation of those who teach mathematics at elementary and secondary levels and a restructuring of the systems that motivate and reward university mathematicians. Neither will occur anytime soon.

In the meantime, the future scientists hunker down in classrooms and learn what mathematics they need to do their work. And those in the humanities endure the minimal mathematics instruction required of them, all the while silently vowing to never go near the subject again. The mathematicians go on with their work. And "the two cultures fly apart like galaxies on opposite sides of an expanding universe."

Much as I desire the teaching revolution, *Mathematics in 10 Lessons* is not its manifesto. Instead, the book aims more modestly toward the smoothing over of a small portion of the boundary gap between the cultures. The book stands on the assumption that there exists a significant number of nontechnical readers who—for whatever reason—are ignorant of mathematics and who now wish to set things right. I see the intended audience as being composed of well-educated people who routinely attend concerts, visit museums, and go to the theater.

This is a wide and general audience. And—just as a speaker often works best when he concentrates on a particular member of his audience—an author needs to imagine a specific reader. It is natural to think of a *poet* as being an extreme representative of the nonmathematical N-culture, and I have already, without comment, slipped this notion into the introduction. And a *physicist* comes to mind as truly a representative of the M-culture. While this may be unfair stereotyping to the small collection of poets who have significant mathematical knowledge and interest, I will continue to think of my typical reader as being a poet. If I can reach a poet with mathematics, I tell myself, I can reach anyone. Moreover, we will later see that there exist striking analogies between the practice and characteristics of poetry and those of mathematics. Poets are closer to mathematicians than they know.

Mathematics in 10 Lessons stands on the premise that humanists can learn mathematics if they wish, and if the subject is properly presented to them. A proper presentation—this book asserts—flows gracefully from the pedagogical principle that requires the careful and leisurely examination of a few basic mathematical notions. In the mathematical field, a single notion precisely learned is worth a textbook of partially digested material. If the objects of the study are properly chosen, and if they are completely mastered, then they provide a springboard for the future study of more complicated ideas—notions that when extended lead directly to powerful and modern mathematics.

The book begins with truly elementary ideas. But elementary ideas are not easy ideas and their mastering requires effort. Understood ideas lead to understanding. Misunderstood ideas lead nowhere.

After the introduction, the book is divided into ten chapters. The idea is to consider each chapter a mathematics *lesson* where each lesson deals, more or less, with a particular mathematical topic. Some of the lessons are more extensive than others but each ranges in length between what can be presented in a single mathematics class period and what can be presented in three such periods. Of course, one cannot learn all of mathematics in ten lessons, nor in a thousand lessons. But the ten lessons given here are basic lessons. They move slowly and are written for a nontechnical audience. Learn them well and they will take you to whatever mathematics you want, or need, to know.

I hope that this book also appeals to scientists. Properly read, it should increase the scientist's aesthetic sensitivity toward mathematics and thus bring the scientist closer to the poet, to whom truth and beauty are always the same thing. But the book is truly written for poets. My earlier book showed the poet the mathematical bridge between the two cultures. *Mathematics in 10 Lessons* shows the way across.

Chapter 1

TRUTH AND BEAUTY

Like Macbeth's air-drawn dagger, mathematics lives in the mind. Mathematical objects—things like numbers, equations, matrices— are abstract and imagined. They do not belong to the real world. Certainly, mathematical symbols can be written on pages and printed in books that are indeed real-world objects. But the subject itself comes from somewhere deep in the mind. Mathematics is made of pure thought, as are air-drawn daggers. It is not made of ink. Mathematicians make it out of airy nothing. And, as poets do, they give it form:

> . . . as imagination bodies forth
> The forms of things unknown, the poet's pen
> Turns them to shapes and gives to airy nothing
> A local habitation and a name.[1]

As I write I have an illustration before me—perhaps by N. C. Wyeth—of a unicorn standing with his spiraled horn in the air and pale moonlight on his silver flanks. Beyond him lies water and stone towers that shine above the trees like Camelot. It is a lovely picture and in it the silver creature seems truly fabled. But it is only a picture. Yet it is as close as any of us will come to unicorns.

Go look. Search the dark forest. Stand still as a tree. Wait beside the bright pool until the moon wanes and you turn cold as stone. You will see no unicorns. Unicorns do not live on this earth. Neither does mathematics.

As you search, you may find a ledger left behind in some abandoned campsite. On the first page you may see faded mathematical symbols including, let's say, the number 6. But you cannot conclude from this that "sixes" exist in the real world. To do so would be analogous to concluding that unicorns exist because N. C. Wyeth once drew a picture of one of them.

What you have found, on the old ledger, is a picture of "the number six"—a symbol that represents a mathematical idea. All those who know elementary mathematics share this idea. It connotes to them other commonly shared ideas.

For example, they know that "six" is the name of the natural number (natural numbers are the numbers: 1, 2, 3, 4, 5, 6, 7, . . .) that follows five and precedes seven. Also, they are aware of certain arithmetical properties of 6 such that $6 = 1 + 2 + 3$, and $6 = 2 \cdot 3$. (Here the "dot" represents multiplication. This is one of the symbols we will use for this operation. Nowhere in this book do we indicate multiplication with the "×," which is commonly used in elementary school. "×" looks too much like "x," which we will use for something else.)

Thus, the number six is an *idea* that stands apart from the world of reality, as do unicorns or daggers made of air. Moreover, we can extend this discussion from the number six to any mathematical notion. Each mathematical notion—no matter whether it is as simple as that of a positive integer or as complicated as a topological space—is an abstraction and lives in the world of ideas.

As we proceed, it will become increasingly important to formalize this distinction. Think of the two worlds—the world of reality and the world of mathematics—as existing side by side. They can be represented schematically in various ways. Figure 1 shows each world as a simple rectangle, the real world on the left and the mathematical world on the right. Real-world objects—the things that live there—are just what you think: poems and people, palaces and plasterboard, all those things you can see or touch. *Real things.*

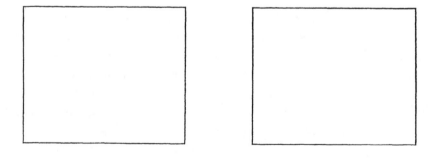

Real World   Mathematical World
Figure 1: Worlds real and abstract

The other, the mathematical world, contains only ideas. But they are special ideas and they have names like *number* or *function* or *inequality*. These are *mathematical ideas* and they have properties that they inherit from simpler mathematical notions and *they can be manipulated and extended—according to the rules of mathematics—to produce more complicated ideas.* Part of our purpose in this book is to come to grips with the mathematical world and to understand some of the fundamental ideas that live in it. This will come as our story unfolds. What we want to understand now is the existence of the mathematical world and its separateness from the world of reality.

It is in this sense of separateness, of abstraction, that mathematics is similar to Wyeth's unicorn or to Macbeth's dagger of the mind. All are mere constructs, composed of nothing but pieces of pure thought. But there is a significant sense in which mathematics differs from the others. The dagger comes from the heat-oppressed brain of a man who would be king and who stands on the bloody edge of murder. The unicorn derives from Dark Age myth and from the tales of magic told on long nights around leaping fires. Each has its place in literature and in the unfolding of civilization. But neither is *necessary*. The world would be pretty much as it is if neither the dagger nor the unicorn had been conceived—not as rich a world, but still *the world*. Malory and Shakespeare are important to me and I prefer a world containing all their ideas. I do not like to imagine a world without the concepts of unicorn or that of Macbeth's dagger. But I can imagine such a world. And I know it would be not much changed.

On the other hand, it is inconceivable to imagine a world without mathematics. (Without mathematics there would exist no science beyond mere description and categorization. The world would be absent the great predictive and explanatory powers provided by mathematics. No part of such a world could be free of opinion, of dogma, or of wishful thinking.) The two worlds of figure 1 are fundamentally entwined like the branches of adjacent vines. They are *essential* for one another. The real world gives us people who create mathematics. The mathematical world gives us *truth*.

CREATION

A lot has been slipped into the final two sentences of the above. I've mentioned the two philosophically fundamental concepts of *creation* and *truth*. We need to take a closer look at each of them.

As we have seen, the people who pursue research in mathematics are called *mathematicians*. And when they engage in research they refer to the process as "doing mathematics." To "do mathematics" means to set down on paper new mathematics: mathematics that has heretofore not existed. Ordinarily, this new mathematics becomes the content of a research paper that is published in a specialized journal that exists solely for the purpose of receiving such papers. The research paper then becomes accessible to other mathematicians who may read it and be stimulated to do further research.

Only mathematicians do mathematics. In fact, in certain circles this is taken as the definition of the term "mathematician." "A mathematician," they say, "is a person who does mathematics," "and mathematics is what mathematicians do"—obviously circular definitions, but completely intelligible to those in the business. One of my purposes is to make it intelligible for all of us.

Notice that this "definition" of mathematician says nothing about teaching mathematics. Most mathematicians are employed in universities and they do teach mathematics. But that is different from *doing* mathematics. A research mathematician is likely also to be a teacher of mathematics. But not always. Moreover, there are many, many people who teach mathematics at some level or another who do not pursue mathematics research, and who may even—like most of the educated public—be unaware that such research exists. Teaching is one thing, research is another.

So, when someone introduces himself as a mathematician, the proper response may not be "what kind of mathematics do you teach?" but rather "what kind of mathematics do you *do*?" The mathematician—being then properly impressed and gratified with your knowledge of his field—will answer something like: "differential equations," or "topology," or "complex analysis," thereby giving you the name of the particular subfield of mathematics in which he works. (There are, incidentally, more than fifty subfields of mathematics identified by *Mathematical Reviews*, a major publication of the American Mathematical Society.)[2]

A central question arises quickly from all this: "What exactly does a mathematician do when he sets down new mathematics? Does he *discover* this mathematics or does he *create* it?

On the surface these two points of view seem clearly distinct. Does a mathematician *find* new mathematics or does he simply make it up? The first point of view—that of *discovery*—is known as the "Platonist," or the "absolutist" viewpoint. The other—that of *creation*—is referred to as the "constructivist" position. The Platonist view is the more passive, seeing mathematicians more or less as discoverers of an existing mathematical reality. The constructivists see mathematicians as makers, as creators of a mathematical world that without them would not exist. But as the question is pursued at deeper philosophical levels, the two views tend to fuse and the mathematician emerges as creator sometimes and as discoverer at other times. An individual mathematician often feels he is inventing some or other specific piece of mathematics while simultaneously uncovering bits of mathematical reality through his work as a whole. However, the fundamental question remains: "Is mathematics created or discovered?"

The question is philosophically complicated and has yet to be settled to anyone's satisfaction. We are not about to settle it here. Nor should we try. But I want to make two points.

The first is that my own view is the constructivist view: *mathematics is created.* This is the view that will prevail throughout this book and I shall consistently refer to mathematicians as creators of mathematics. I consider the mathematical world (as figure 1 shows) as a world of ideas that is separate from the world of reality. (As I have already noted, the two worlds are entwined. But the separate branches do not touch.) Mathematicians live in the real world and they create the objects that live in the mathematical world.

The second point is that mathematicians themselves are divided on this issue and my position is the minority view. An informal survey of colleagues at various institutions has shown a ratio of about eight to two in favor of *discovery* over *creation*. Most mathematicians—I am convinced—believe they discover, not create, mathematics. And there are some nonmathematicians on the discovery side as well. An example is Martin Gardner, the distinguished mathematical expositor, who wrote in a recent book review:

> Not only is the Universe mathematically structured, it is
> made entirely of mathematics.[3]

If Gardner is correct then not only is mathematics discovered, but the two worlds of figure 1 actually merge. Everything in the universe turns into mathematics. I do not believe him correct but we must take Mr. Gardner seriously. He has thought deeply and written beautifully about mathematics for many years and he stands squarely in the Platonist camp.

I remain unconvinced. If the Platonists are to prevail, they must come to grips with a fundamental, and yet unanswered, question:

> If mathematics is discovered, then it must already exist.
> Who created it?

Obviously, you can discover only that which already exists. The Platonists must then provide for mathematics a creator other than a mathematician. Or else, they are obliged to provide a process by which mathematics came to be. Those who believe the world to be made of mathematics should first convert the cosmologists. For the cosmologists—who worry about the creation of the universe—clearly are looking at the wrong problem. What the Platonist position requires is not a theory for the origin of the physical universe, but rather one for the origin of mathematics.

We could, of course, put our faith in a revised version of Genesis 1: *In the beginning God created mathematics.* Otherwise, we need a new cosmology. We need a big bang theory for mathematics.

I have none to offer. Neither have the Platonists. I'll settle for the constructivist viewpoint. Mathematicians create mathematics.

TRUTH

At first, it seems paradoxical to assert, as did the mathematician Alfred Renyi, "One can know more about things which do not exist than about things which do exist."[4] Yet this is the view that must prevail if we are to find truth in the mathematical world rather than in the real world. For the mathematical world consists only of ideas and abstraction, things that certainly do not exist in the sense that a chair or a tree exists.

Renyi is correct. The truth or falsity of any statement about the real world ultimately depends on its confirmation or denial by examination in the light of observation. Ultimately, we must go to the real world and look.

Consider, for example, this statement: *all crows are black.* Here we have a simple declarative sentence that asserts the existence of some property of the real world, namely, each fowl that belongs to the collection of birds called crows is black in color. It seems obvious that the assertion is true or else it is false. Either there exists a nonblack crow or there does not. If such a bird exists, the statement is false. Otherwise, it is true.

However, there seems to be little chance of determining—with complete certainty—which of these alternatives holds true. A process might have to be developed whereby every single crow on the face of the earth could be examined for the presence or absence of nonblack feathers. Of course, you might be lucky and find early on a nonblack crow. If you do, the process ends, and you know the statement is false.

But you might not locate such a bird. (I've never seen one.) If not, then you continue. You go on observing and recording the blackness of the world's crows one by one until, somehow or other, you have examined them all. The process seems endless and unpractical. How will you know when you are done?

Moreover, there are serious verification problems that go beyond the technical difficulties associated with the examination process. One is the philosophical question of the reliability of observation itself.

All of us—at some time or another—have encountered a situation in which one or another of our senses has failed us—as did Macbeth's eyesight when he saw the floating dagger with its handle toward his hand. Coffee smells like tea when I have a cold. Glass feels like marble underwater. And once from my window, I saw walking in the rain with his old familiar stride a colleague who had been dead for two years. I know—we all know—that our senses are not always completely reliable. How then can we be certain about black crows? We cannot.

Of course, we can come close. If we do the observation carefully and scientifically, we can eliminate most of the sources of potential error. We can be pretty sure. But being pretty sure is not being certain. And it is not what we mean by truth. You do not find truth by *looking*.

Moreover, our discussion about crows extends to any assertion regarding the real world. In order to test the hypothesis that a given real-world object has some particular property, a scientist *observes* the object. The observation may be complicated and it may require the use of sophisticated laboratory equipment, but it is observation nevertheless. You cannot be certain of what you see. When a friend claims that the house across the lake is white, your proper response is: "It seems white. On this side."

Because real-world validity depends ultimately on observation, we must look somewhere else for truth. The real world is the place for wonderful things—sunsets and flowers and poetry. It surely is the place for love. But truth lives somewhere else. Truth lives in the mathematical world.

LOGIC

At its heart mathematics consists of the examination of certain kinds of statements called *implications*. The implications are assertions of a particular type about the abstract objects that make up the mathematical world. Each implication falls formally into three parts. The first part is itself a statement called the *hypothesis*. The third part is another statement called the *conclusion*. Connecting the hypothesis and the conclusion is a symbol that stands for *logical implication*. In this book, we will use the symbol $\Rightarrow$ for this.

So, if we denote the hypothesis by p and the conclusion by q, our implication may be denoted by $p \Rightarrow q$. We read $p \Rightarrow q$ as "p implies q."

Let's notice in passing that the introduction of "$\Rightarrow$" constitutes our first encounter with a new mathematical symbol. We will subsequently see many others. In fact, the book will soon seem crammed with symbols. But this is the nature of things and there is no avoiding it. Mathematics is written in symbols and to learn the subject means to partly learn its symbols. You must learn its *way of saying*. The trick is to master the symbols easily, one by one. The mastering will bring you satisfaction at first. Then felicity.

Part of the mastering of the symbol "$\Rightarrow$" requires nothing more than repetitive practice in writing it and saying it. You should invent simple implications such as:

$$x = 6 \Rightarrow 2x = 12$$

and

$$y = 0 \Rightarrow y + 2 = 6,$$

and say them aloud:

"x equals 6 implies $2x$ equals 12"

and

"y equals 0 implies y plus 2 equals 6"

until the association of "$\Rightarrow$" with the word "implies" becomes automatic. You should also notice the great difference of these two implications. The first implication is *valid*, the second is *not valid.*

We say that the implication $p \Rightarrow q$ is *valid* if q is true whenever p is true. Thus $p \Rightarrow q$ may be thought of as an assertion that "q follows from p" or equivalently as "if p then q." An implication that is known to be valid and that has some measure of significance is often called a *theorem*. To *prove* a theorem means to establish—using the laws of logic and the rules of mathematics—the validity of the implication contained in the statement of the theorem. To do research in mathematics essentially means to create, and prove, theorems.

One can also consider implications about real-world objects. For example, our earlier statement, "all crows are black," can be reformulated as: "if C is a crow, then C is black." This takes the form of the implication:

$$C \text{ is a crow} \Rightarrow C \text{ is black.}$$

The discussion about observation and the fallibility of sense-perception can then be summarized by saying: "it is not possible to establish the validity of this real-world implication."

Of course, the validity of many real-world implications is clear. An example is

$$C \text{ is a crow} \Rightarrow C \text{ is a bird.}$$

Another is

$$D \text{ is a dog} \Rightarrow D \text{ is an animal.}$$

Since these last two statements merely restate part of the definition of "crow" or "dog," neither implication possesses much significance. But the validity of each is transparent.

Associated with the implication $p \Rightarrow q$ is the reverse implication $q \Rightarrow p$, which is called its *converse*. For example, the converse of

$$C \text{ is a crow} \Rightarrow C \text{ is a bird}$$

is

$$C \text{ is a bird} \Rightarrow C \text{ is a crow.}$$

Here, it is clear that—although the original implication is valid—the converse is not. (Not all birds are crows.) Thus, in general, the validity of $q \Rightarrow q$ does not yield the validity of the converse $q \Rightarrow p$.

Sometimes we will want to write simultaneously both an implication and its converse. We then write $p \Leftrightarrow q$ and read this as "p if and only if q." An obvious example is

$$D \text{ is the day after Sunday} \Leftrightarrow D \text{ is Monday}$$

because this just says:

"D is the day after Sunday if and only if D is Monday."

Other standard terminology associated with implications must be learned. If $p \Rightarrow$ q is valid, we say that p is *sufficient* for q or that q is *necessary* for p. If $p \Leftrightarrow q$ holds, we say that "p is *necessary and sufficient* for q" (or that "q is necessary and sufficient for p").

Therefore,

"C is a crow" is *sufficient* for "C is a bird,"

"*C* is a bird" is *necessary* for "*C* is a crow,"

and

"*D* is Monday" is *necessary and sufficient* for

"*D* is the day after Sunday."

In summary, our terminology is:

> $p \Rightarrow q$ is translated as "*p* implies *q*" or
> "if *p*, then *q*" or
> "*q* follows from *p*" or
> "*p* is sufficient for *q*" or
> "*q* is necessary for *p*."

Also,

> $p \Leftrightarrow q$ is translated as "*p* implies *q*" and "*q* implies *p*" or
> "*q* if and only if *p*" or
> "*p* is necessary and sufficient for *q*" or
> "*q* is necessary and sufficient for *p*."

This is a surfeit of terminology, but all of it appears in the mathematical literature sufficiently often that it must be mastered. I will use it frequently enough in this book that you will find the mastering natural and easy.

The main point is that mathematics consists of implications. In mathematics *p* and *q* are statements about certain mathematical objects. You suppose *p* to be true and then—using logic and the rules of mathematics—you *deduce* statement *q*, thereby proving the theorem: $p \Rightarrow q$. At the outset *q* may be unknown to you, a mathematical secret awaiting discovery. In this case, you might think of the implication symbol as an arrow pointing from the known to the unknown.

Robert Frost says we circle around real-world secrets.[5] But in mathematics, truth sits in the middle. And you go at it head-on. The implication symbol points the way.

COMPOUND STATEMENTS

Logicians talk about "compound statements" and they develop formal methods for constructing and evaluating compound statements from simpler statements. The simple statements are combined by means of what are called *connectives.* One of the ultimate objectives of symbolic logic is to construct a kind of logical calculus that allows you to manipulate the symbols so that the truth or falsity of any compound statement becomes transparent whenever the truth or falsity of each simpler component is known.

Although it is not our purpose to present here even a subcourse in symbolic logic, we will from time to time want to examine in some detail a particular compound statement. For the moment, we will concentrate on the particular compound logical statement: $p \Rightarrow q$.

Here, the component statements are "the statement p" and "the statement q." And the connective between them is "the implication symbol $\Rightarrow$." Thus, "C is a crow" is a simple statement. Another is "C is a bird." Putting them together with the symbol "$\Rightarrow$" yields the (valid) compound statement:

$$C \text{ is a crow} \Rightarrow C \text{ is a bird.}$$

Logicians examine compound statements with a device called a *truth table.* This device is simply a schematic method for setting down systematically the truth or falsity of a compound statement in terms of the truth or falsity of its component statements. The rules for manipulation of the logical connectives in any compound statement—when properly applied—produce the truth values of the statement in much the same way as algebraic manipulation of component symbols allows a mathematician (sometimes) to find the solution of complicated equations.

These logical rules (sometimes called *rules of inference*) are themselves developed from simpler rules, which come from rules even simpler, and so on until the logician comes back to the fundamental *definitions* of the truth values of the basic connectives (from which all compound statements are ultimately composed).

So far, we have considered only one connective, the symbol "$\Rightarrow$." (The "if and only" connective is just implication in both directions.) And we have examined it only in the light of simple real-world statements whose truth or falsity seemed intuitively obvious. But real truth—as we have argued—does not live in the real world. Our purpose is to examine the concept of mathematical truth (and its magical association with the real world). Consequently, if it is true—as we have also asserted—that the notion of implication lives at the heart of mathematics, then certainly we must set down at the outset a formal method for determining the validity of a *mathematical* implication $p \Rightarrow q$. The easy way to do this is to write out the truth table that *defines* the truth value of $p \Rightarrow q$. When this has been done, we may apply the truth table to determine the value of the implication under consideration, whether it is mathematical or not.

In the preceding, we have spoken somewhat casually of the *validity* of the implication $p \Rightarrow q$. This word has been used in order not to confuse the truth or falsity of the compound statement with that of the component statements p and q. But the meaning is the same. Any mathematical statement— simple or compound—is either true or else it is false. Mathematics operates in a *two-valued* logic and these values are *true* or *false*. (There are some deep foundational questions that deal with the issue of whether or not it is always possible to determine whether a particular statement is true, but this is another matter. For us, any mathematical statement is true or else it is false. No other possibilities exist.)

Therefore, when we say that the compound statement $p \Rightarrow q$ is *valid*, we simply mean that it is true. Similarly, to say it is *invalid* means the same as saying it is false. And we determine the truth value of $p \Rightarrow q$ by looking at the statement set down in its truth table, which defines this value. The table is shown in figure 2.

p	q	$p \Rightarrow q$
T	T	T
T	F	F
F	T	T
F	F	T

Figure 2: Truth table for $p \Rightarrow q$ (p implies q)

In figure 2 the first column lists the possible truth values of p. These values are indicated as T or F, which, of course, means *true* or *false*, respectively. The second column gives similar values for the statement q. The third column shows the corresponding values for the compound statement $p \Rightarrow q$. The table has four rows because it must show all possible combinations of T and F for the statements p and q. (It is easy to see that there are four such combinations. Later, we will examine a general method for doing this sort of counting.) Thus, the first row of the table in figure 2 tells us that *when p is true and q is true, then $p \Rightarrow q$ is true*. The second row shows that $p \Rightarrow q$ is false if p is true and q is false.

The last two lines of figure 2 deserve special attention since they are somewhat nonintuitive. They tell us that $p \Rightarrow q$ is true whenever statement p is false. Thus

"the moon is made of green cheese" $\Rightarrow$ "crows are black"

is a true implication. So is

"the moon is made of green cheese" $\Rightarrow$ "crows are blue."

(Both implications are *true* because each hypothesis is *false*. See figure 2.)

At the beginning of our discussion of $p \Rightarrow q$, we referred to p as the *hypothesis* and to q as the *conclusion* of the compound statement. So, the table in figure 2 tells us that the implication $p \Rightarrow q$ is always true except in the single case in which the hypothesis is true and the conclusion is false. Moreover, it is crucial that we understand that there is no question of the truth or falsity of what the table in figure 2 tells us about $p \Rightarrow q$. The truth table *defines* the implication. In any particular case, we may worry about the truth value of the hypothesis or that of the conclusion but, once these are known, the table provides the truth value of $p \Rightarrow q$.

I must point out here that this notion of $p \Rightarrow q$ is not completely standard. Many logicians would consider figure 2 as the truth table for the statement $p \rightarrow q$, which is called a "conditional." This enables them to distinguish between a *formal compounding* of statements as defined in figure 2 and a *relation* between statements. An *implication*, to these logicians, is a relation $p \Rightarrow q$, which means that "q logically follows from p." More precisely, they mean "$p \Rightarrow q$ is true if and only if the conditional $p \rightarrow q$ is true in all cases where p is logically true."

This distinction seems esoteric for a beginning discussion and I shall not make it. Its main advantage is that it allows the logicians to avoid the apparent paradoxes of such implications as:

$$\text{walruses can fly} \Rightarrow \text{pigs have wings.}$$

The implication is true (as figure 2 shows) because the hypothesis is false. But the logicians are troubled because there is no logical connection between hypothesis and conclusion.

None of this shall trouble us and we will make no distinction between implication and conditional. Ultimately, *both* our p's and q's will be statements about mathematical objects. We will not deal with silly implications like:

$$\text{if it rains today, then } 2 + 2 = 5.$$

Nor will we concern ourselves with whether pigs have wings or why the sea is boiling hot.

In the case of a mathematical implication $p \Rightarrow q$, the hypothesis usually asserts that some given mathematical object possesses some property or other. The conclusion then claims that some object then has some other property. In the *proof* of the implication, the truth of the hypothesis is *assumed*. Then the validity of the conclusion is *deduced* by some process or other that usually involves the proper manipulation of mathematical symbols. A mathematician normally does not care whether or not p is true. He is concerned only with the validity of the statement:

$$\text{If } p \text{ is true, then } q \text{ is true.}$$

Moreover, the mathematical object to which the hypothesis applies normally is a general representative of a class of mathematical objects and, thus, is not a *particular thing*. These observations are what led Bertrand Russell to his often quoted (and generally misunderstood) remark:

> Thus mathematics may be defined as the subject in which
> we never know what we are talking about, nor whether
> what we are saying is true.[6]

The significance of Russell's remark and its exact meaning will become increasingly clear as we work our way (beginning in chapter 2) through some mathematical implications. But, before we begin, we need to talk a bit more *about* mathematics and we need a slight extension of the logical apparatus that will allow us to manipulate its objects.

THREE LITTLE WORDS

Let's examine the words: *and, or, not.*

The compound statement $p \wedge q$ is read "p and q." The statement is *defined* by the truth table given in figure 3. Hence $p \wedge q$ is true only when both p and q are true. If either is false, then $p \wedge q$ is false. (Notice that p and q are statements but neither is a hypothesis nor a conclusion. These terms apply only to implications.)

p	q	$p \wedge q$
T	T	T
T	F	F
F	T	F
F	F	F

Figure 3: Truth table for $p \wedge q$ (p and q)

The compound statement $p \vee q$ is read "p or q." This statement is *defined* by the truth table shown in figure 4. Thus, $p \vee q$ is true except in the single case where both p and q are false.

p	*q*	*p* ∨ *q*
T	T	T
T	F	T
F	T	T
F	F	F

Figure 4: Truth table for *p* ∨ *q* (*p* or *q*)

For example,

"crows are birds" ∨ "pigs can fly"

is true because one of the statements is true. (See figure 4.) But

"crows are birds" ∧ "pigs can fly"

is false because one of the statements is false. (See figure 3.)

Notice that the connective ∧ has the meaning we usually associate with the word "and." When we say "*p and q*" we mean *both* must hold. This is exactly what we mean when we say that *p* ∧ *q* is true: *p* is true and *q* is true. You may safely think "and" whenever you see ∧. But you need to be more careful with identifying ∨ with the usual meaning of "or."

In ordinary speech, the word "or" is often ambiguous. If I say "I will watch the night's starred face or I will sit in the moonlight," you understand I may well do both. But when I tell you "the world will end in fire or it will end in ice," you know that both possibilities cannot hold. And sometimes you cannot tell. The sentence "I will run or write tomorrow" may well mean I am uncertain. I may do one or the other; perhaps I will do both.

But there is no ambiguity about *p* ∨ *q*. This statement is defined by the truth table in figure 4. The connective is *inclusive* because *p* ∨ *q* allows for *p* or *q* or *both*. (Logicians have an exclusive version of whose truth table is the same as that of figure 4, except that the compound statement is false whenever both components are true.) When we come to the formulation of mathematical statements using the word "or," we must take care about which version we are using.

If p is a statement then, the new statement ~p is called "the negation of p" and is read "not p." The truth table for ~p is shown in figure 5. Thus p is true only when p is false. If p is "crows are black," then ~p means "crows are not black."

p	~p
T	F
F	T

Figure 5: Truth table for ~p (not p)

We can use negation and the connectives ∧, ∨, and ⇒ to build compound statements from simpler ones. As an example consider the statement $p \wedge \sim q$ (read "p and not q"). We can build the truth table for $p \wedge \sim q$ by proceeding methodically from the truth values of p and q and making use of the truth table for the connective ∧ and that of negation ~. The procedure is shown in figure 6. The result is that $p \wedge \sim q$ is true only in the case where p is true and q is false.

p	q	~q	$p \wedge \sim q$
T	T	F	F
T	F	T	T
F	T	F	F
F	F	T	F

Figure 6: Truth table for $p \wedge \sim q$ (p and not q)

The third column of the table in figure 6 is technically not a part of the truth table for $p \wedge \sim q$. It is shown to indicate the method of construction.

Look, for example, at the third row of the table. Here p is false and q is true. Thus, $\sim q$ is false. Then we see from the truth table for $r \wedge t$ (just figure 3 with p replaced by r and q replaced by t) that $p \wedge \sim q$ is false. The other rows are reconstructed analogously.

The next example deserves special attention:

EXAMPLE 1. Consider the statement $\sim p \vee q$. We construct the truth table in the same systematic way as before: first we write a column for the statement p and then one for q. (We do p first only because of convention.) We place in these columns all four possibilities of T, F choices for p and q. Then we make a column for $\sim p$, which consists of the opposite of the values in the column for p. Finally, we use the truth table for the connective to complete a final column for $\sim p \vee q$. The result is shown in figure 7.

p	q	$\sim p$	$\sim p \vee q$
T	T	F	T
T	F	F	F
F	T	T	T
F	F	T	T

Figure 7: Truth table for $\sim p \vee q$ (not p or q)

Once again, the third column does not properly belong to the truth table for $\sim p \vee q$. I include it as a helpful step toward the construction of the final column that lists the truth values of $\sim p \vee q$ corresponding to those of p and q. Now delete the third column from the table in figure 7 and compare the reduced table to that of figure 2, which is the *defining* truth table for the implication $p \Rightarrow q$. You see that the two tables are identical. Because of this, we say that the two statements, $p \Rightarrow q$ and $\sim p \vee q$, are *equivalent*.

In particular,

$$\text{"}C \text{ is a crow"} \Rightarrow \text{"}C \text{ is a bird"}$$

is equivalent to

"*C* is not a crow" or "*C* is a bird."

Another statement that is equivalent to both $p \Rightarrow q$ and $\sim p \vee q$ is the statement: $\sim q \Rightarrow \sim p$.

EXAMPLE 2. The statement $\sim q \Rightarrow \sim p$ is called the *contrapositive* of $p \Rightarrow q$. Its truth table appears in figure 8. The construction proceeds as usual from left to right across the rows. Consider, for example, the third row of the table. Here p is false and q is true. Thus $\sim p$ is true and $\sim q$ is false. Then $\sim q \Rightarrow \sim p$ is true since any implication $r \Rightarrow t$ is true whenever the hypothesis r is false. (See figure 2.)

p	q	$\sim p$	$\sim q$	$\sim q \Rightarrow \sim p$
T	T	F	F	T
T	F	F	T	F
F	T	T	F	T
F	F	T	T	T

Figure 8: Truth table for $\sim q \Rightarrow \sim p$ (not q implies not p)

Now compare the truth table in figure 8 (delete the construction columns 3 and 4) with the tables in figures 2 and 7. You see they are identical. Hence, the three statements, $p \Rightarrow q$, $\sim p \vee q$, and $\sim q \Rightarrow \sim p$, are equivalent.

You must be careful not to confuse the contrapositive of $p \Rightarrow q$ with its converse.

EXAMPLE 3. The statement $p \Rightarrow q$ is called the *converse* of $q \Rightarrow p$. Its truth table comes easily from figure 2 via an appropriate interchange of letters. Better yet, we can set it down immediately if we remember that $r \Rightarrow t$ is always true except in the single case where r is true and t is false. The table is shown in figure 9.

p	*q*	*q* ⇒ *p*
T	T	T
T	F	T
F	T	F
F	F	T

Figure 9: Truth table for *q* ⇒ *p* (*q* implies *p*)

In a sense, nothing new comes from writing down the table for $q \Rightarrow p$ since we could obtain it from an interchange of the roles of *p* and *q* wherever they appear in figure 2. However, we are interested here in the absence of symmetry between $p \Rightarrow q$ and the converse $q \Rightarrow p$. So when we compare the tables, we want to keep the same first columns and then compare the final column, which gives the truth value for $p \Rightarrow q$ and $q \Rightarrow p$, respectively. When this is done, we see the tables are not the same. (For example, the third row of figure 2 tells us that $p \Rightarrow q$ is true whenever *p* is false and *q* is true. But the third row of figure 9 shows that, under the same conditions on *p* and *q*, the converse $q \Rightarrow p$ is false.) Therefore, an implication and its converse are not equivalent.

This does not mean that $p \Rightarrow q$ and the converse $q \Rightarrow p$ cannot both be true. They can be and we have already seen an example in

"*D* is the day after Sunday" ⇒ "*D* is Monday."

In fact, we used this implication in both directions as an example of $p \Leftrightarrow q$. Let's look at the truth table for the "double implication."

EXAMPLE 4. The statement $p \Leftrightarrow q$ means (by definition) $(p \Rightarrow q) \wedge (q \Rightarrow p)$. We can construct the truth table for this compound statement by looking at the tables for $p \Rightarrow q$, $q \Rightarrow p$, and $r \wedge t$. These are, respectively, given in figure 2, figure 9, and figure 3, with symbols appropriately changed. The parentheses tell us—in the final step—to think of *r* as $p \Rightarrow q$ and of *t* as $q \Rightarrow p$. The table is shown in figure 10. (Look at the construction of the second row: p is true, q is false, so that $p \Rightarrow q$ is false. But $q \Rightarrow p$ is true. Then $(p \Rightarrow q) \wedge (q \Rightarrow p)$ is false.)

p	q	$p \Rightarrow q$	$q \Rightarrow p$	$(p \Rightarrow q) \wedge (q \Rightarrow p)$
T	T	T	T	T
T	F	F	T	F
F	T	T	F	F
F	F	T	T	T

Figure 10: Truth table for $p \Leftrightarrow q$ (p if and only if q)

Earlier, I said that $p \Leftrightarrow q$ is read "p if and only if q." The "if and only if" language should mean that $p \Leftrightarrow q$ is true exactly when p and q have the same truth value. You see from figure 10 that this is precisely what occurs. The compound statement $p \Leftrightarrow q$ is true exactly when p and q are both true or they are both false.

Just for fun, let's put these ideas together to examine a more complicated statement.

EXAMPLE 5. In example 3 we defined the double implication $p \Leftrightarrow q$ by the compound statement $(p \Rightarrow q) \wedge (q \Rightarrow p)$. Consequently, the first statement, $p \Leftrightarrow q$, should hold if and only if the second statement, $(p \Rightarrow q) \wedge (q \Rightarrow p)$, holds. We can hook these two statements together with the connective to form the statement

$$(p \Leftrightarrow q) \Leftrightarrow [(p \Rightarrow q) \wedge (q \Rightarrow p)].$$

Admittedly, this statement is a mess. To write it we need both parentheses and square brackets in order to know the order of the operations. But the compound statement inside the square brackets defines the first statement on the left $(p \Leftrightarrow q)$. And, since $(p \Leftrightarrow q)$ and its *defining statement* are connected by the "if and only if" connective, it is natural to expect that the complete statement *will always be true*. Is it? The only way to be sure is to write out the truth table. (It would be a good exercise for you to try this on your own. You need only proceed methodically using the known truth tables and following the indicated order of operations. First, set down the values for *p* and *q*. Then look at $p \Leftrightarrow q$. Next, consider $p \Rightarrow q$ and then $q \Rightarrow p$. Then join these two statements with the connective $\wedge$. Finally, join this last result with ($p \Leftrightarrow q$) using the connective $\Leftrightarrow$. You are done. You should have before you a column of all *T*'s.) The constructed table is given in figure 11.

p	q	$p \Leftrightarrow q$	$p \Rightarrow q$	$q \Rightarrow p$	$(p \Rightarrow q) \wedge (q \Rightarrow p)$	$(p \Leftrightarrow q) \Leftrightarrow [(p \Rightarrow q) \wedge (q \Rightarrow p)]$
T	T	T	T	T	T	T
T	F	F	F	T	F	T
F	T	F	T	F	F	T
F	F	T	T	T	T	T

Figure 11: Truth table for $(p \Leftrightarrow q) \Leftrightarrow [(p \Rightarrow q) \wedge (q \Rightarrow p)]$
(*p* if and only if *q*) if and only if [(*p* implies *q*) and (*q* implies *p*)])

A statement, like that of example 5, which is always true, is called a *tautology*. A simpler example is:

"*C* is a crow or *C* is not a crow."

Let's close this brief foray into symbolic logic with a story often told in mathematical circles.

A scientist wishes to test the claim that all crows are black. He grabs binoculars and a clipboard and heads for the fields. Once there, he begins to observe birds.

"There's a crow," he says to himself, "and it's black."

He meticulously records this observation by making a mark on paper. And he continues the process, observing and marking, through a long day in the hot sun.

"Another black crow," he says. "And another, and another."

By day's end he's seen three hundred crows, all of them black. He comes home tired but pleased with his work. He has, he thinks, accumulated considerable evidence to support the claim "all crows are black."

Has he? A mathematician would proceed this way: the claim "all crows are black" is the same as the implication

$$\text{"C is a crow"} \Rightarrow \text{"C is black."}$$

This, in turn, is logically equivalent to the contrapositive

$$\text{"C is not black"} \Rightarrow \text{"C is not a crow."}$$

If you want to test the former, you may as well test the latter. They are the same.

So while the scientist heads for the fields, the mathematician leans back in his chair and looks around his office.

"There's a yellow pencil," he says. "It is not black and it is not a crow. I'll mark this down."

The mathematician then has a single observation to support the contrapositive and thus to support the original implication since they are the same. He continues to look around his office. He finds a blue book that is not a crow and then a white tablet, also not a crow. In no time at all he locates three hundred nonblack items that are not crows. All of this, he marks down. Without leaving his chair, he accumulates the same amount of supporting evidence as did the scientist in the field.

Or did he? You decide.

MODELS

In the next chapter we will begin to apply our discussions of symbolic logic to implications of the form $p \Rightarrow q$, where p and q are *mathematical statements*. Often, our procedure will be as follows:

First, we set down a statement p that makes some assertion about an object, or a collection of objects, which live in the mathematical world. We will *assume* that statement *p* is true. Thus, statement *p* becomes a *hypothesis*. Next, we use the laws of logic and the rules of mathematics to *deduce* from *p* the correctness of another statement *q*. The deductive process—which usually involves the appropriate manipulation of mathematical symbols—provides the proof of the implication $p \Rightarrow q$.

Often, the conclusion *q* is not known in advance but is *discovered* through the deductive process. When this happens, statement *q* may express some completely unexpected—and perhaps bizarre—property of the mathematical objects under consideration. But unexpected or not, proving the implications enables us to say with mathematical certainty: "if *p* is true, then *q* is true." The implication then becomes an example of truth.

We are speaking here of mathematical truth, not truth about the real world. And since our fundamental assumption (see figure 1) is that the real world and the mathematical world are separate—one being a world of *things* and the other being a world of *ideas*—it would seem that this notion of truth is artificial and maybe even sterile. What possible utility can it have? Indeed, if the real world and the mathematical world are disjointed, what did Galileo have in mind when he told us centuries ago:

Nature's great book is written in mathematical symbols.

One way to reconcile this apparent contradiction is as follows: consider a piece of the real world that you wish to examine. This "piece" is indicated by the shaded region in figure 12. Perhaps you want to study the motion of a falling raindrop, or the spread of a new influenza virus.

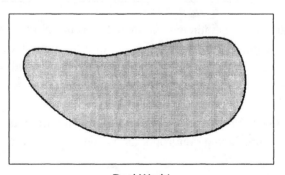

Real World
Figure 12: A piece of the real world

Maybe the piece of the real world that interests you is the San Andreas Fault because you want to predict the frequency of California earthquakes. It does not matter. In any case, we denote the part of the real world under consideration by the shaded area in figure 12.

Next, we consider a copy of this portion of the real world over in the mathematical world. This process I call *abstraction* and it is shown in figure 13. The copy produced in the mathematical world (also shown in figure 13) is often known as a *mathematical model*. This model lives in the mathematical world, and therefore is composed entirely of mathematical objects like numbers or equations.

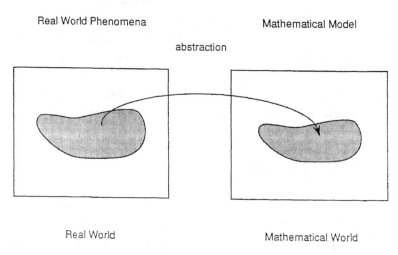

Real World Phenomena   Mathematical Model

abstraction

Real World   Mathematical World

Figure 13: Mathematical modeling

Consequently, the mathematical model does not imitate the real world in the ordinary sense. A mathematical model of a falling raindrop is not a photograph of a raindrop. Nor is it a realistic painting. Instead, the model captures the features of the raindrop that the mathematician considers essential to the situation under consideration.

Let's suppose the situation has to do with motion. Our mathematician wants to determine how a raindrop falls. Specifically, he wants the answer to the following question:

If a raindrop starts to fall at time $t = 0$, what is its velocity v at an arbitrary time t?

(Do not be put off by the strangeness of the phrase "starts to fall at time $t = 0$." No doubt a real-world raindrop forms and begins to descend at a specific hour like 6:15 AM or 10:42 PM. But, in the model, it is convenient to think of the starting time as $t = 0$, which amounts only to a rescaling of the clockface. The mathematician then concerns himself with the behavior of the raindrop one second later, two seconds later, or more generally, t seconds later.)

A raindrop, like all objects near the surface of the earth, is pulled toward the earth's center by gravitational force. This force equals the product of two quantities: the *mass* of the object (which depends on the object) and a constant known as the *acceleration due to gravity*. The number produced by this multiplication is simply the weight of the object. Thus, your own weight—that which shows on your bathroom scale—is the product mg, where m denotes your mass and g denotes the acceleration produced by gravity. (In ordinary units, the constant g approximately equals 32 feet per second per second. For now just think intuitively of velocity as speed and of acceleration as increase or decrease of speed. These terms will be rigorously defined later.)

What interests the mathematician are the forces acting on the raindrop. According to Isaac Newton's famous second law of motion, these forces determine the raindrop's acceleration, but acceleration represents the rate of change of velocity and the mathematician knows how to recapture velocity if acceleration is known. Similarly, he knows how to produce the raindrop's position once he knows its velocity, which is simply the rate of change of position. (The necessary mathematical technique is called *integration* and belongs to that part of the mathematical world called *calculus*. We will come to this technique later on.)

If we now let m denote the mass of the raindrop and we again let g denote the acceleration produced by gravity, then the force pulling the raindrop toward the earth equals mg, the weight of the raindrop. (Remember that juxtaposition of symbols, like mg, means they are multiplied together.)

But, there must be some other forces acting on the raindrop. Otherwise, the raindrop would fall like a stone and the mathematician believes it should not. A stone—the mathematician knows—is essentially unaffected by air resistance. Compared with its weight, the resistive force of the atmosphere to the motion of the stone is small enough to be neglected. Not so, however, with the raindrop. As the raindrop falls, the "pushing back" of the atmosphere against it must be taken into consideration as the mathematical model is constructed. Thus, the model must show two forces acting on the raindrop: the weight *mg* pulling it down, and another force: the air resistance retarding the motion. What is this second force? Or more properly: *what shall we assume it to be?*

Here, the mathematician must use his insight and intuition. He may want to make use of laboratory data that have been obtained from actual measurements of real-world falling raindrops. Maybe he goes out in the rain and observes. But, sooner or later, he must invent the resistive force as he creates the model. As the mathematician constructs the raindrop model, he simultaneously sets down the hypothesis p for a mathematical implication $p \Rightarrow q$, whose conclusion he does not yet know. But the hypothesis must include the resistive force. Ultimately, he must *decide* what this force has to be.

Let's decide for him. It seems reasonable to assume that the resistive force will vary directly with the raindrop's velocity. That is, the faster the raindrop falls, the more strongly the atmosphere resists the motion. Let's *assume* this is the case.

CV

mg

Figure 14: Forces on a raindrop

This assumption can be expressed mathematically by writing the resistive force as cv where v denotes the velocity of the raindrop at an arbitrary time t and c denotes a (presently unknown) constant. (The constant c is independent of time while the raindrop's velocity v will vary with time. Often the dependence of v on t is indicated by writing $v = v(t)$, which is read "vee equals vee of tee.") Figure 14 shows the raindrop and the forces acting on it. When these forces are combined with Newton's second law of motion, the model jumps out.

The second law of motion says that the net force acting on a moving body equals the product of the mass of the body and the body's acceleration. If we denote the force by F, the second law becomes

$$F = ma$$

where m is the mass and a the acceleration.

In the case of the raindrop, the net force pulling it toward the earth is (see figure 14) $F = mg - cv$. The acceleration of the raindrop can be written as $a = dv/dt$, where the latter quantity represents a calculus notion called "the derivative of v with respect to t." (This is a sophisticated calculus notion, which is carefully explained in chapter 8.) When these quantities are substituted into Newton's second law above, we obtain: $mg - cv = m \cdot dv/dt$. (For typesetting purposes, we often write the fraction $\frac{a}{b}$ as a/b. So dv/dt means $\frac{dv}{dt}$.

The dot between m and $\frac{dv}{dt}$ denotes multipliction.) When both sides of this equation are divided by the mass m, it becomes: $dv/dt = g - (c/m)v$. The mathematician's assumption that the raindrop *starts to fall* at $t = 0$ means that the initial velocity is zero, that is, $v(0) = 0$. This provides the model with what is called the *initial condition*. Thus the model can be completely written with the two equations:

(M)
$$\frac{dv}{dt} = g - \frac{c}{m}v \text{ and}$$
$$v(0) = 0.$$

This is the mathematician's model. Equations (M) paint a picture of a falling raindrop in the mathematical world. These equations show a falling raindrop as it looks to a mathematician.

The mathematician can do more with a model than just look. The first of the two equations shown in the model is called a *differential equation*. Using techniques of mathematical analysis (only calculus is needed in this situation), the mathematician can solve this equation for the raindrop's unknown velocity. Moreover, the expression he obtains shows v as a specific function of time t. The solution turns out to be:

(S)
$$v = \frac{gm}{c}\left(1 - e^{\frac{-ct}{m}}\right).$$

At the moment we are not concerned with the details of the calculation that led to the solution (S). Nor will we bother with the exact meaning of all the symbols involved in the solution. (The number e that appears in (S) will turn up later in the book. We will see that it is approximately equal to 2.7 and learn of its practical and aesthetic importance as one of the truly fundamental constants in all of mathematics.) But we want to make clear the *significance* of the solution.

Our mathematician began with a part of the real world he wanted to study: in this case a falling raindrop. Next he produced—over in the mathematical world—model (M) of this piece of real-world phenomena. Now the mathematician leaves the real world and, using pure mathematics, he manipulates the model to obtain truths he did not know before. Here, the model consisted of a single differential equation together with an initial condition. The differential equation expresses a condition that the rate of change of the raindrop's velocity must satisfy. The mathematician solves the differential equation and produces from the very air of the mathematical world the solution (S). From equation (S) he can compute the velocity of the raindrop at any given time by simply substituting numbers into the right-hand side of equation (S).

Let's emphasize again that this step in the process *takes place completely in the mathematical world*. This is *pure mathematics*. It involves only the manipulation of mathematical objects. If we consider the equations of (M) as a mathematical statement M and the solution in (S) as another statement S, then our mathematician has proved the theorem:

$$M \Rightarrow S.$$

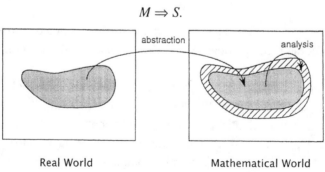

abstraction   analysis

Real World   Mathematical World
Figure 15: Manipulating the model

This step is called *analysis* and is shown in figure 15. The mathematical results produced by the analysis appear as the diagonally shaded area in the figure. In our example, this consists of the expression for v given in (S). In general, it consists of whatever we learn from the mathematical manipulation that we did not know before. This new area represents mathematical truth.

But what good is it? What the mathematician has obtained evidently applies only to mathematical raindrops, to raindrops of the mind. Certainly, the mathematician has learned something he did not know at the start. In the beginning he knew only the forces on the raindrop and its initial velocity. Now he knows the raindrop's velocity at an arbitrary time—certainly a significant accomplishment. He obtained this information solely by means of pure mathematics, through the appropriate manipulation of mathematical symbols. I have no doubt that he has found truth, in the strictest sense of the word. But it is truth about an imaginary raindrop; truth that appeared at the end of a correct chain of mathematical argument. This truth consists of nothing more than the proof of the mathematical implication: $M \Rightarrow S$. Can this abstract and seemingly artificial processes tell us anything about real-world raindrops? It seems unlikely.

For this dry rain is rain made of mathematics, not of water. Its properties are only those set down by the mathematician in the model and those produced by him in the manipulation of the model. But in the real world, rain wets as fire burns. The mathematical raindrop dampens no ground. It falls forever in the perfectly resistive atmosphere of the mathematical world. Real raindrops fall through blowing wind. They wet your hair, splash the window, drip from leaves. Surely the one can have nothing to do with the other.

But they do. The mathematical truths tell us everything about the model. They also tell us about the world. Real raindrops fall pretty much as the mathematics tell them to fall. The model's truths are *applicable* to the real world exactly as Galileo promised us centuries ago. This applicability holds in the raindrop model and it holds in many other mathematical models. Amazingly, in the bright light of the real world the model becomes reality. The mathematician completes the process by moving back to the real world (see figure 16) and *applying* the result of the mathematical analysis to the actual phenomena under consideration. The model comes home from the mind as does the sailor from the sea. And it works.

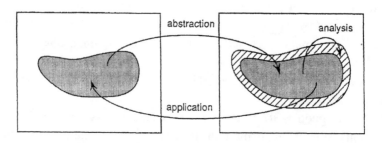

Real World   Mathematical World

Figure 16: The applied mathematics process

The renowned physicist and Nobel laureate Eugene P. Wigner called this "the unreasonable effectiveness of mathematics."[7] I agree. The great utility of mathematics is indeed unreasonable. And it is much more. In the sunlight, mathematics metamorphoses and becomes magic.

Chapter 2

ONE, TWO, THREE, . . . , INFINITY

Let's begin by noticing that *this chapter bears a misleading title*. The title seems to indicate that the list of numbers that begins with 1, 2, 3 terminates at something called *infinity*. But this list does not terminate—not at infinity or anywhere else. As we shall see, it just goes on and on.

If you want to refer to the "whole numbers" from 1 to 6, you might say: "Consider the numbers 1, 2, 3, 4, 5, 6." No problem here, the set of numbers under consideration is sufficiently small that the numbers can simply be written in a list. But if you wanted to look at the numbers from 1 to 1,000, it might be inconvenient to write them all. The list would be too long. In this case, a mathematician would say: "Consider the numbers 1, 2, 3, . . . , 1,000."

The acceptable convention is that the three dots mean "and so on" so that the symbol "1, 2, 3, . . . , 1,000" tells us to look at the integers that begin with 1 and continue consecutively until we reach 1,000. More generally, the symbol "1, 2, 3, . . . , n" tells us to consider the integers that begin with 1 and continue until we reach the fixed, but apparently unknown integer n. (With this convention, the above list 1, 2, 3, 4, 5, 6 could be replaced by the slightly more abbreviated symbol 1, 2, 3, . . 6.)

Consequently, the chapter's title, "One, Two, Three, . . . , Infinity," would indicate—according to the convention—that we want to consider the list of numbers that begins with the number 1 and concludes with a number called "infinity." However, it is crucial—now, at the beginning of our study of mathematics—we understand that *no such concluding number exists. There is no number called—"infinity," or whatever—that exceeds each of the "whole numbers."* The phrase "one, two, three, . . . , infinity" has a nice ring and even a bit of rhyme. The physicist George Gamow found it appealing and used it as the title of his well-known book.[1] I like it too, as you see. But it has no real mathematical meaning.

To be sure, several true mathematical notions of "infinity" exist and we will come to one of them shortly. But—when we do—we will find that it is not a number in the sense that we will understand "number" in this book. Simply put, *there is no number whose name is "infinity."*

The numbers whose names we know best are, of course, the numbers 1, 2, 3, ... We may remember the names of certain other numbers—like "pi" or "the square root of 2"—but they are less familiar and certainly less natural than the counting numbers 1, 2, 3, ... "Six" is better known than "pi." Ask anyone.

In fact, the numbers 1, 2, 3, . . . are called *natural numbers*. These are the numbers first encountered by schoolchildren and they constitute the material of which the entire subject called arithmetic is constructed. Arithmetic—as taught in our schools—consists in the transmission (often by rote), from teacher to pupil, of the allowable rules for proper manipulation of these numbers. That is, in the study of arithmetic you learn how to combine the natural numbers with one another by means of operations called addition and multiplication, and by the inverse operations of subtraction and division. (We will come to these operations shortly.)

If we were to begin the study of arithmetic properly, we would first set down a minimum set of fundamental assumptions, called *axioms*. Then from these axioms—and appropriate definitions—we would *derive* the allowable manipulative rules for the natural numbers. This process would consist of stating and proving theorems about natural numbers. At each stage, the theorems would be proved using only the assumed axioms and whatever theorems had been previously established. And it is exactly this process—not the rote memorization of rules and procedures—that truly makes up the subject called mathematics.

Mathematics, as we have seen, consists entirely of the demonstration of validity of certain statements in the form of "*p* implies *q*." What you do in mathematics—at the highest levels—is produce *proof* that a particular mathematical statement *q* is true whenever some other statement *p* is true. And the proof you write will itself be a collection of mathematical statements each following from its predecessor in a clear and distinct chain of argument that can be, if need be, traced all the way back to the axioms on which the subject is founded.

Part of our work in this book will be to make clear the meaning of this interpretation of mathematics and to establish—or partially establish—a number of particular "*p* implies *q*" statements; statements that, since proved, have become *theorems*. Before we are finished, we will state and prove a number of these theorems.

Of course, a complete development of even a part of mathematics would require us to trace the logic all the way back to the axioms on which the mathematics is founded. This we will not do for two reasons. First, such a development requires a level of rigor that—in my view—is inappropriate for an introductory view of mathematics. I do not want us to become bogged down in the mire of logical subtleties and tedious detail that would be needed to establish, for example, that there exists no natural number between 6 and 7. Such things we simply take for granted. Second, we will avoid complete rigor because, to the novice, it has about it an atmosphere of deadly sterility. At the lofty levels of the practicing mathematician, complete rigor defines for mathematics what Bertrand Russell called "supreme beauty" and makes it "the most intellectual of the arts."[2] But this stern beauty shows itself only to those who look patiently and hard, and for a long time. The novice will not see it. Nor does he—at the beginning—want to see or need to see it yet.

The newcomer wants to see what mathematics is about and why the collection of strange people called mathematicians pursue it with such religious fervor. And he wants to see this as quickly as possible. But, in mathematics, speed is incompatible with rigor. To be rigorous means necessarily to move slowly. Our goal is to move apace through a certain subset of mathematical topics. In order to do this, we will eschew complete rigor.

But not all rigor. We will, as I have said, prove a certain number of theorems. But the proofs will depend on certain assumptions we specifically make about the mathematical objects under consideration, not on the founding axioms.

Nevertheless, it is important for us to know that the axioms exist and that our work could be pushed back to them by any number of practicing mathematicians. And it is this "pushing back"—this setting down of an exact set of logical statements, which goes from the fundamental axioms to the particular theorem under consideration—that determines the nature of mathematical truth. The statement: "The square of an odd number is odd" is a mathematical truth precisely because a proof can be written that leads all the way back to the fundamental axioms.

We will not go to such a far place. We will take the existence of the natural numbers for granted as did the mathematician Leopold Kronecker, who wrote in 1887:

God made the natural numbers; all else is the work of man.[3]

COUNTING

Children count by pointing. Ask a child how many apples there are on the table. Watch. He will lean forward in his highchair, extend an index finger, point to an apple, and, as best he can, say the word "one." Then he moves his finger to another apple and says "two." Another repetition, the spoken word "three," and he is done. There are three apples on the table.

Or are there? We need to examine carefully what the child was trying to do (instinctively, no doubt) and how he might go wrong.

Let's suppose the child knows the names of the first few natural numbers and let's assume that there are indeed three apples on the table. Moreover, let's assume the apples are distinct enough from one another—either through size and skin color or position on the table—that we can tell them one from another. Consequently, we can denote the apples by three distinct symbols, say A, B, and C.

If the child is correctly counting, he points sequentially from one apple to another and then to the other. We may as well suppose he points first to A, then to B, and finally to C. When he points to apple A and speaks the word "one," our child is tacitly *assigning* to A the natural number 1. Put another way, the child *pairs* the apple A with the number 1. A mathematician would denote this pair by the symbol $(A, 1)$. Similarly, by the finger-pointing method the child assigns 2 to apple B and 3 to apple C. When he is done, he has created the three pairs $(A, 1)$, $(B, 2)$, and $(C, 3)$.

Probably, the child does not think of pairs. More likely, he tacitly thinks of "naming" each apple on the table. A becomes apple 1, B apple 2, and C apple 3. But no matter what the language, the process is clear: the child has produced a *correspondence* between the collection of apples on the table and the natural numbers 1, 2, and 3 in such a manner that each apple is associated with exactly one of the numbers 1, 2, 3. It is the establishment of this correspondence that constitutes the act of counting apples.

There are two obvious ways in which the counting can go wrong. First, the child could omit one or more of the apples. For example, he might point to apple A, say the number "one," point to B, say "two," and then point to A again and speak the number "three." In this case, the child has arrived at the natural number 3 that truly represents the number of apples on the table. Nevertheless, he has counted incorrectly as those of us who are watching will surely point out. What the child has done is form the pairs $(A, 1)$, $(B, 2)$, and $(A, 3)$. He has properly used each of the numbers 1, 2, 3 in the pairing but he has not used all the apples. He has not established a correspondence between the set of apples and the set of numbers that associates exactly one of the numbers with each of the apples.

The child could also miscount by giving the same number name to two (or more) of the apples. He might, for example, point to 1, say "one," and then point to B and say the same thing. If this happens, the child might as well point to apple C and say "two." In this case, the child has created the pairs $(A, 1)$, $(B, 1)$ and $(C, 2)$. He has miscounted because the correspondence he has established does not associate an apple with each of the numbers 1, 2, 3.

To properly count the apples, the child must—by pointing or whatever—*establish a correspondence that associates with each apple* A, B, C *one and only one of the numbers* 1, 2, 3. He failed in the first instance because he did not use all the apples in the pairing. In the second case, he miscounted because he did not use all the numbers. (Notice, that in the first failure he used one of the apples twice in the pairing and in the second miscounting he used one of the numbers more than once.)

Obviously, the child could miscount in many other ways. (He might, for example, speak numbers randomly like "eight," "five," and "ten.") But these two are the erroneous apple-counting methods of present interest. And, to be truthful, we are not much interested in apples anyway. We want to learn to speak mathematics and apples are not mathematical objects. What we must do is paint a picture of the apple situation over in the mathematical world of figure 13. We do not need a Cezanne still life. We need a mathematical model of the collection of apples.

A fundamental mathematical notion is that of a *set*. Indeed, in a certain sense the concept of set is the most basic notion in the subject. After Georg Cantor produced his formal study of sets back in 1874[4] a school of mathematicians arose that claims the set concept forms the logical foundation for all of mathematics. The members of this school—sometimes called the "logicists"—argue convincingly that mathematics properly begins with a careful delineation of exactly what is, and what is not, a set.

Once again, it is not our purpose (nor are we prepared) to dive into such logical depths. We must come to grips with the notion of set, but we will come to it naively. For us, *a set is simply a collection of objects.* Such a collection will be considered properly defined whenever we have some rule that allows us—at least theoretically—to determine whether or not a given object belongs to the set.

We will, from time to time and mostly through examples, talk freely about sets of real-world objects. Already such a set has been tacitly introduced when we spoke in the above counting example of the "set of apples." We might also speak of the "set of people in the room" or "set of trees in the backyard" since in each case we presumably can determine the objects that belong to each of these collections. But mainly we will deal with sets of mathematical objects and we will establish certain properties that these sets possess. Thus we will talk about sets in the following way: "the set of natural numbers that are not greater than three." (The members of this set are, of course, the numbers 1, 2, 3.)

If the members of a set are explicitly known, then the set can be most easily described by simply writing them down. When this is done, it is conventional to name the set with a single symbol (usually a capital letter) and to enclose the members of the set in curly brackets. For example, we might say:

(1)   Let $S = \{1, 2, 3\}$.

This means that *S is a set whose members are the numbers* 1, 2, *and* 3. (The curly brackets tell you S is a set and the symbols inside tell you its members.) Equation (1) therefore defines a particular set whose name is S.

Now let's make a great leap. Consider the statement:

(2)   Let S be a set.

Our leap consists of the gap between equation (1), which defines a particular set S, and sentence (2), which assigns the name S to an arbitrary set. (We cannot conclude, of course, that the two sets named S are equal. The context of our discussion tells us that (1) and (2) are independent of each other. The point I want to make about generality of statements could be introduced by replacing (2) with: "let *T* be a set." However, we need to become familiar with the common mathematical practice of using the same general symbol to represent different particular objects.) The *S* in sentence (2) has no given particular properties. If we want to talk about it, we can assume only that it possesses those properties that are common to all sets. Thus, given the little we know now about sets, we can assume only that *S* is a collection of objects. But we know not what objects, or how many.

Please do not underestimate the significance of the generality of statement (2). Understanding it—and similar statements about mathematical objects—constitutes the beginning of the understanding of abstraction and, hence, of mathematics. As we go forward with our study we will encounter many similar statements: "let *n* be a natural number," "let *C* be a circle," or "let *T* be a triangle." In each case, the symbol represents a general object of the type indicated. Without further information we cannot conclude, for instance, that $n = 6$, *C* has radius 1, or *T* is isosceles. If *S* is an arbitrary set given by (2), then it contains certain objects as members. (Unless we are told otherwise, we will always assume *S* to be a set of mathematical objects.) We need a symbol for *an arbitrary member* of *S*. A common choice for this member is *x*. When this holds, we write

$$x \in S.$$

The symbol $\in$ stand for "epsilon," the fifth letter of the Greek alphabet. But the collective symbol $x \in S$ is read: "*x* belongs to *S*" or equivalently as "*x* is a member of *S*."

If we want to indicate that an object *x* does not belong to *S* (equivalently, *x* is not a member of *S*) we write

$$x \notin S.$$

So, if we go back to the set $S = \{1, 2, 3\}$ given by (1) we see that

$$1 \in S,$$
$$2 \in S, \text{ and}$$
$$3 \in S$$

are correct statements while

$$6 \in S$$

is false. (But $6 \notin S$ is true.)

Often, it will be convenient (and perhaps necessary) to describe a set in some manner other than by simply listing its elements in curly brackets. In fact, this has already been done when the set $S = \{1, 2, 3\}$ was referred to as "the set of natural numbers that are not greater than 3." A more precise (and more general) method for this is as follows: suppose $P(x)$ represents some statement about the object x. Then we will use the notation

$$S = \{x: P(x)\}$$

to mean "S is the set of all x such that $P(x)$ is true."

For example, $P(x)$ might be the statement "x is a natural number that is not greater than 3." Then, if

$$S = \{x: \text{"x is a natural number not greater than 3"}\}$$

it follows that

$$S = \{1, 2, 3\}.$$

As a second example of these two kinds of notation for sets, consider the set E, which consists of all the *even natural numbers*. The elements of E then consist of the numbers 2, 4, 6, 8, 10, ... So we could write the set E as

$$E = \{2, 4, 6, 8, 10, 12, \ldots\}$$

where the three dots tell us that the list inside the curly brackets continues forever in the same fashion. Alternately, we might notice that a natural number n is even if and only if $n = 2m$, where m is some other natural number. ($m = 0$ gives $n = 0$, $m = 1$ gives $n = 2$, $m = 2$ gives $n = 4$, $m = 3$ gives $n = 6$, and so forth.) Thus we may write

(3) $$E = \{n: n = 2m, m \text{ is a natural number}\}.$$

Before we go further, we need a symbol for the set of natural numbers. A standard notation for this set is $\mathbb{N}$, which we will consistently use throughout the remainder of this book. Hence, by definition of $\mathbb{N}$:

$$\mathbb{N} = \{1, 2, 3, 4, 5, 6, \ldots\}.$$

We may then use this notation to write the set of even integers given by (3) as:

$$E = \{n\colon n = 2m, m \in \mathbb{N}\}$$

or even more succinctly as

$$E = \{2m\colon m \in \mathbb{N}\}.$$

Next we must slightly enlarge our intuitive concept of a set. We have taken a set to mean simply a "collection of objects" and, tacitly, we have assumed that each given set contains some objects (maybe just a single object like the set $\{6\}$). But we must allow for the possibility that a set may contain no objects. An example would be

$$T = \{n\colon n \in E, n = 7\}.$$

Here, E denotes the set of even numbers given by (3) and the comma inside the curly brackets is read as "and." So the set T describes the set of all even integers that are equal to 7. But there are no such numbers since 7 is not even. Thus T contains no elements.

The set that contains no elements is called the *empty set* and is denoted by the symbol $\emptyset$. (This is not phi, the twenty-first letter of the Greek alphabet, but is a version of the fifteenth English letter, O.) Hence, the statement $x \in \emptyset$ is false for every x while $x \notin \emptyset$ is true for every x. If S is not the empty set, we say that S is nonempty.

The next concept—that of *subset of a set*—is important enough to define formally.

DEFINITION 1. Let S and T be sets. We call T a subset of S, if and only if $x \in T \Rightarrow x \in S$, and we write $T \subset S$. If $T \subset S$ and $T \neq S$, we say that T is a proper subset of S.

(A general remark about definitions needs to be made at this time: *all definitions are if and only if statements*. All of us—even mathematicians—are sometimes careless and write a definition as only an "if statement." But we understand it to be if and only if. Thus, if we write:

A triangle is called isosceles if it has two equal angles

we mean:

A triangle is called isosceles if and only if it has two equal angles.)

Thus, to say that T is a subset of S means that whenever x belongs to T then x belongs to S, that is, each member of T is also a member of S. In terms of the contrapositive of the implication in the definition we have: T is a subset of S if and only if each element not in S is also not in T.

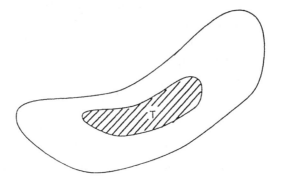

Figure 17: *T* is a subset of *S*

Often it is helpful to denote sets graphically as regions of a plane. When this is done (see figure 17) we must show T as being inside S whenever T is a subset of S.

We need to pay careful attention to the following:

REMARK. The empty set is a subset of every set. That is, if S is any set, then $\emptyset \subset S$.

There are at least two ways to see that the remark is valid. First, notice by definition that $\emptyset \subset S$ if and only if each element $\emptyset$ is also an element of

S. Since Ø contains no elements, this condition is automatically satisfied. Alternately, notice that to say Ø "is not a subset of S" means that there exists some element in Ø that is not in S. Since there are no elements in S, this cannot be true. Hence it is true that $Ø \subset S$.

Notice that *a set is always a subset of itself*, that is, $S \subset S$ for any set S. ("$x \in S$ implies $x \in S$" is obviously true.) Thus a nonempty set S always has at least two distinct subsets Ø and S.

If we now let S denote the specific set $S = \{1, 2, 3\}$ given by (1), it is not difficult to see that S has exactly eight subsets. They are

(4)   Ø, $\{1\}$, $\{2\}$, $\{3\}$, $\{1, 2\}$, $\{1, 3\}$, $\{2, 3\}$, $\{1, 2, 3\}$,

the last set being, of course, S itself. Each of the others is a proper subset of S. (In this list of subsets of S we did not include $\{2, 1\}$ because this set is identical with $\{1, 2\}$ since they both contain exactly the same elements. Similarly, $\{2, 3\} = \{3, 2\}$ and $\{1, 3\} = \{3, 1\}$. Including both $\{1, 2\}$ and $\{2, 1\}$ in the list of subsets of $\{1, 2, 3\}$ would be improper as it would be to write Beethoven's name twice in a list of history's ten greatest composers.)

The list (4) tells us that it is proper to write $\{1\} \subset \{1, 2, 3\}$ or $\{2\} \subset \{1, 2, 3\}$, and similar inclusions for the other subsets of S. Also, $1 \in \{1, 2, 3\}$. However, $1 \subset \{1, 2, 3\}$ is *false*. (1 is not a subset of $\{1, 2, 3\}$.) In fact, if x is an element of an arbitrary set T then $\{x\}$ is a subset of T so that $\{x\} \subset T$ is true. However, it is *improper* to write $x \subset T$. (But $x \in T$ is correct.)

In the earlier "apple-counting" example we saw our child establish, by the method of finger pointing, a correspondence between the set $S = \{1, 2, 3\}$ and the set of apples $T = \{a, b, c\}$, which associated with each element of S an element of T and vice versa. (The change of notation to lowercase letters for the apples will become clear momentarily.)

DEFINITION 2. Let T and S be nonempty sets. Let C be a rule or a correspondence that associates with each element $x \in T$ a unique element $y \in S$. Moreover, suppose each element $y \in S$ is associated with a unique element x of T by means of C. Then we write $y = C(x)$ and call C a one-to-one correspondence between T and S. (The equation $y = C(x)$ is read "y equals C of x.")

Thus a one-to-one correspondence C between T and S can be considered as a pairing of elements, x in T with $y = C(x)$ in S, such that each element in T is paired with exactly one element in S and vice versa. In the finger-pointing example, the child established a one-to-one correspondence C between $T = \{a, b, c\}$ and $S = \{1, 2, 3\}$ by means of the pairings $(a, 1)$, $(b, 2)$, $(c, 3)$. In terms of the $y = C(x)$ notation this is the same as writing: $C(a) = 1$, $C(b) = 2$, and $C(c) = 3$. (Here, you see the reason for the change to lowercase letters. Had the change not been made, the last equation would have awkwardly read $C(C) = 3$.) The establishment of this one-to-one correspondence allowed the child to intuitively conclude that the sets T and S contained the same number of elements. This allowed him to assert that T contained three elements because he knows that the set $\{1, 2, 3\}$ contains three elements. So, there are three apples on the table.

The child counts this way. So do we all. There are eighteen books on the top shelf of the bookcase nearest my desk. I just now determined this by pointing and counting. I went book by book: 1, 2, 3, 4, . . . , 17, 18. Consequently, I established a one-to-one correspondence between the set of books on the shelf and the subset of natural numbers $\{1, 2, 3, . . . , 18\}$. Thus the two sets contain the same number of elements. Eighteen.

We can generalize this counting process by means of the following:

DEFINITION 3. Let T and S be nonempty sets. We say that T and S contain the same number of elements if and only if there exists a one-to-one correspondence between T and S.

Let's illustrate this concept with two children, Arthur and Brent, who are twins as young as the grass is green and who cannot count. They sit at the kitchen table, across from one another. It is afternoon treat time. Their father places a portion of jellybeans in front of each child. Arthur and Brent eye the piles suspiciously. And they eye each other. Who has more jellybeans?

Arthur stretches forth his hand and takes a single jellybean. Brent, simultaneously, does the same. Each pops the chosen jellybean in his mouth. They chew slowly and swallow together at the same moment. Then they repeat the process: select a single jellybean, place it in the mouth, chew, and swallow. They reach, chew, and swallow in unison like mirror images of one another, each staring at the other across the table like a bird of prey. One by one the beans disappear and the piles dwindle. Finally, a single bean remains in front of each child. The hard stares turn to smiles. Each child takes his last bean and wolfs it down. They have consumed—each child knows—exactly the same number of jellybeans.

If we think of Arthur's original portion of jellybeans as being the set A and Brent's as the set B, then what the children have done with their cautious process of selecting and eating is to establish a pairing of the elements of A and B in such a way as to produce a one-to-one correspondence between the elements of A and B. Consequently, A and B have the same number of elements according to definition 3.

Had the children already learned sufficiently the names of many natural numbers, the pairing of the jellybeans one with another would have been unnecessary. Arthur could simply count his jellybeans. He would do this by some kind of sorting process that would establish a one-to-one correspondence between his set A of jellybeans and a subset (say) $\{1, 2, 3, \ldots, m\}$ of the natural numbers. Then Arthur knows he has exactly m jellybeans. Brent can do the same thing and see that he also has m beans. Then they can eat leisurely, knowing that neither of them has more than the other.

In fact, the reason children are taught to count is precisely so that they can avoid the (perhaps) tedious task of pairing objects whenever they want to determine whether or not two sets of objects contain the same number of elements. If the objects are properly counted, then they are placed in correspondence with some subset of the natural numbers of the form $\{1, 2, 3, \ldots, k\}$. The set just counted then contains k objects. The natural numbers provide us with a set of objects whose names we know. We carry this set in our heads and use it as the set with which we pair the elements of any other set whose size we want to determine.

However, the direct pairing method used by Arthur and Brent has considerable value and we need to examine it more carefully. In the first place, had Arthur run out of jellybeans before Brent, he would have concluded instantly, and probably loudly, that Brent had begun with a larger portion. Brent would have arrived at the same conclusion had his portion been exhausted first. Second, the pairing method, that is, the method of establishing (or trying to establish) a one-to-one correspondence between sets A and B, has meaning whether or not it is possible to count the elements of either of these sets. Arthur and Brent could not count their sets because they have not yet learned the natural numbers. Let them grow a bit older and give them some instruction, then counting jellybeans will no longer be a problem. Arthur had placed before him a finite set of jellybeans. So had Brent. When they grow older, they will count the elements in these sets with ease and, thus, quickly determine they contain the same number of jellybeans. But definition 3 tells us how to determine whether or not two sets have the same number of elements with no mention of the natural numbers. Definition 3 is applicable even if, in some ideal children's world, Arthur and Brent each were given infinitely many jellybeans.

INFINITE SETS

DEFINITION 4. Let S be a set. Then S is called finite if $S = \emptyset$ or if there exists a one-to-one correspondence between S and a subset $\{1, 2, 3, \ldots, m\}$ of natural numbers.

When S is a finite set, we may speak about the "number of elements in S" and we denote this quantity by $n(S)$. If $n(S) = m$, then S may be placed in a one-to-one correspondence with $1, 2, 3, \ldots, m$ and it would be appropriate to write S as $S = \{x_1, x_2, \ldots, x_m\}$ where the x's represent the elements of S. If S is the empty set, then it contains no elements, that is, $n(\emptyset) = 0$.

As examples of this notation, consider $A = \{a, b, c\}$, and $B = \{1, 2, 3, \ldots, 35\}$. Then $n(A) = 3$ and $n(B) = 35$. If we let T denote the set of all trees growing at this moment in the state of Tennessee, then it is clear that T is a finite set so that $n(T)$ has meaning. But we have no idea as to the exact value of $n(T)$. (No doubt it is quite large.)

The situation becomes different if we look at $\mathbb{N}$, the set of natural numbers. This set fails to satisfy the conditions of definition 4. (Since $1 \in \mathbb{N}$, $\mathbb{N} \neq \emptyset$. If $\mathbb{N}$ could be placed in a one-to-one correspondence with any set of the form $\{1, 2, 3, \ldots, m\}$, then we could list the elements of $\mathbb{N}$ as $x_1, x_2, \ldots x_m$. But then $\mathbb{N} = \{x_1, x_2, \ldots x_m\}$. Now, let y denote the largest of the numbers, $x_1, x_2, \ldots x_m$, and we see that y becomes the largest element in $\mathbb{N}$. But this is impossible since $y + 1$ is a natural number that is greater than y.) Thus $\mathbb{N}$ is not a finite set.

DEFINITION 5. Let S be a set. If S is not finite, then S is called infinite.

(Notice that there is nothing mysterious about this concept of "infinite." Simply put, a set is infinite if it does not satisfy definition 4.)

Another example of an infinite set is the set $E = \{2, 4, 6, 8, \ldots\}$ of even natural numbers. (The proof that E is not finite is analogous to the above argument that $\mathbb{N}$ is an infinite set.) Similarly, the set $K = \{1, 3, 5, 7, \ldots\}$ of odd natural numbers is an infinite set. We also know many examples of geometrically described infinite sets: for example, the number of points on a circle or the number of points on a line segment. (To see that the set of points in a line segment is infinite, just notice that for any two points p and q on a given line segment, the point r_1 midway between p and q is also on the segment. Then the point r_2 midway between p and r_2 is on the segment, and so forth. Thus we can produce an infinite sequence of points $r_1, r_2, r_3, \ldots$ on the segment by repetitions of this process. (See figure 18.)

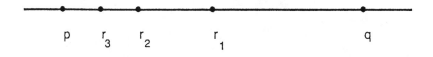

Figure 18: Infinitely many points on a line segment

As we proceed with our study, we will encounter many examples of infinite sets. And, when we meet them, we must be careful about our use of intuition in understanding them. We will often find that intuitive ideas that are perfectly reliable for finite sets fail to be valid for infinite sets. Finite intuition may not blend with an infinite world. The two are often immiscible. An example follows.

Let S be a finite set and let T be a proper subset. Then T contains fewer elements than does S. (That is, $n(T)$ is a smaller number than $n(S)$.) This obvious fact agrees completely with the intuitive real-world axiom that says: "the whole is greater than any of its parts." But the axiom fails for infinite sets.

To see this, consider the set of natural numbers $\mathbb{N}$ and the proper subset of even numbers E. Thus $\mathbb{N} = \{1, 2, 3, 4, \ldots\}$ and $E = \{2, 4, 8, \ldots\}$. We have seen that $E = \{2m: m = 1, 2, 3, 4, \ldots\}$. Now define a one-to-one correspondence C between $\mathbb{N}$ and E by setting $C(m) = 2m$ for each $m = 1, 2, \ldots$. (The correspondence is one to one since each natural number m is paired with exactly one even number $2m$ and vice versa. The pairings produced by this correspondence are described in figure 19.)

We have agreed, I think, that definition 3 (which tells us when two sets contain the same number of elements) gives the intuitively correct answer for finite sets. There is, then, no reason for us not to accept the definition as the correct notion for infinite sets. We have simply paired—by means of the correspondence C—the elements of $\mathbb{N}$ one by one with those of E, exactly as Arthur and Brent would have done had these sets been piles of jellybeans. Consequently, we must conclude that $\mathbb{N}$ and E contain the same number of elements even though E is a proper subset of $\mathbb{N}$. In the infinite world the whole may not always exceed one of its parts.

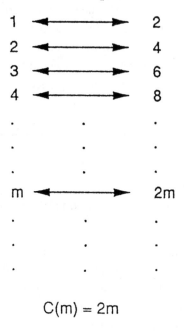

$$C(m) = 2m$$

Figure 19: One-to-one correspondence between the natural numbers and the even numbers

But we must not assume that a one-to-one correspondence between two given infinite sets always exists. So far, the only numbers we have seriously considered are the natural numbers 1, 2, 3, ... Later, we will extend these numbers to the rational numbers and then further to what are known as the real numbers. When this has been done, we will see that the real numbers—denoted by $\mathbb{R}$—form an infinite set of which the natural numbers $\mathbb{N}$ are a proper subset. Thus both $\mathbb{N}$ and $\mathbb{R}$ are infinite sets with $\mathbb{N} \subset \mathbb{R}$ and $\mathbb{N} \neq \mathbb{R}$. But—as we shall see—there is no one-to-one correspondence between $\mathbb{N}$ and $\mathbb{R}$. From this, it is possible to conclude that $\mathbb{R}$ has more elements than does $\mathbb{N}$ even though both are infinite sets. This result gives us another non-intuitive property of infinite sets: some "infinities" are actually larger than other "infinities."

Infinite geometric sets, like circles and line segments, also give rise to nonintuitive results. Consider, for example, the two unequal line segments L_1

and L_2 shown in the first sketch of figure 20. In the second sketch of this figure, the segments have been moved (without stretching or shrinking) so that they are now perpendicular to one another. In the third sketch the end points of the lines have been joined by a dotted line. Now select an arbitrary point q on L_1 and draw a line from q parallel to the dotted line. The intersection of this line with L_2 gives a unique point p on L_2. The procedure works the same way if we begin with an arbitrary point on L_2, draw the dotted line, and obtain a unique point on L_1. It should be clear that this method of projection provides a one-to-one correspondence between the points on L_1 and the point L_2. However, this result holds no matter what the difference in lengths of the two segments. A segment one inch in length contains exactly as many points as does a line thrown between here and a star.

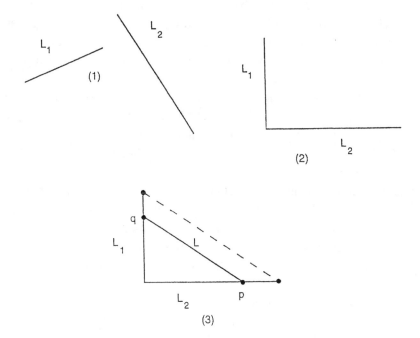

Figure 20: One-to-one correspondence between unequal line segments

Please notice again that nothing mysterious takes place in these analyses. We are dealing with a notion of infinity to be sure, but it reveals itself clearly and simply through definitions 3, 4, and 5. The above results concerning the "size of certain infinities" may be surprising and nonintuitive but they are neither contradictory nor paradoxical. They come to us from the definitions, soft and easy as a summer breeze. And the basic notion on which they stand is that of one-to-one correspondence, which is nothing more than the way kids count jellybeans. Arthur and Brent could tell us that E and $\mathbb{N}$ are the same size. They tell us by reaching and chewing. Wisdom, from the mouths of babes.

RUSSELL'S PARADOX

To reach the modest goals of this book, we can make do with the kind of intuitive and naive treatment of logic and sets that we have so far pursued. We need not worry about the myriad of subtleties that lies at the foundations of these subjects. We will not dive deeply enough in either pool to encounter the rocks on the bottom. Mainly, we need only enough symbolic logic and set theory to allow easy and formal manipulation of the mathematical statements we will come upon. What we have done so far with logic and sets—and that which will follow in the same spirit—will suffice.

Nevertheless, it is appropriate to call attention to the existence of these subtleties, particularly to the class of self-contradictory statements that are known as paradoxes. Let's look quickly at three examples.

A justly famous paradox is credited to the sixth-century Cretan philosopher Epimenides, who said: "Cretans always lie." Is this statement true? If so, then since Epimenides is a Cretan, he must be lying. Thus the statement must be false. Consequently, we have the paradoxical example of a statement with the property that if it is true, then it must be false.

Another, less famous example, came from a personal experience. Once, when my daughter was packing for her first year in college, I walked into her bedroom.

"I have some precollege advice for you," I said.

"I thought you might," she said.

"It's about professors," I said.

"Of course."

"Never," I told her, "never pay attention to anything a professor tells you outside of class."

She looked at me wearily.

"Oh, Father," she said. "You are a professor and we are now outside of class."

I said nothing to that. "Father" is an appellation she uses only when she thinks I have said something particularly stupid. Had I?

A third example comes from set theory. Near the end of the nineteenth century Bertrand Russell (1872–1970), hard at work on the logical foundations of mathematics, came upon a set whose very existence produces a paradox. The set is described below and given the name R in honor of Lord Russell and in recognition that the associated logical difficulties are now collectively known as *Russell's paradox*. But first we need some preliminary remarks.

Ordinarily, a set is not an element of itself. For example, the set of all people attending a concert is not a person, the set of books on my shelf is not a book, the set of keys in my pocket is not a key. On the other hand, there may exist sets that do contain themselves as elements. Perhaps the set of all symbols written in ink is itself a symbol. Maybe the set of all ideas is an idea.

Since the first sets seem more natural than the latter two, let us call the first sets *ordinary* and the other *extraordinary*. Thus a set S is an ordinary set if it does not contain itself as a member. A set T is an extraordinary set if it does contain itself as a member. Hence, S is ordinary if and only if $S \notin S$. T is extraordinary if and only if $T \in T$.

Now let R *denote the set of all ordinary sets.* Is R ordinary or extraordinary? We will check each possibility.

Suppose R is ordinary. Then $R \notin R$ by definition is *ordinary*. But $R \in R$ must hold because R is the set of *all* ordinary sets. Thus $R \notin R \Rightarrow R \in R$.

Now suppose R is extraordinary, that is, $R \in R$. Then $R \notin R$ must hold because R contains *only* extraordinary sets. Thus $R \in R \Rightarrow R \notin R$.

Taken together, these two little arguments prove the strange double implication $R \notin R \Leftrightarrow R \in R$. This is Russell's paradox.

A popular version of Russell's paradox goes like this:

In a village of clean-shaven men there lives a barber. By definition, *the barber shaves those men, and only those men, who do not shave themselves.* Who shaves the barber? If he shaves himself, then he is shaved by the barber. Thus he does not shave himself. On the other hand, if he does not shave himself, then he must be shaved by the barber. Hence the barber shaves himself if and only if he does not shave himself.

Interesting, don't you think? (But still, I wonder what the barber does when he looks into his mirror each morning.)

A way of dealing with Russell's paradox is simply to assert the extraordinary sets do not exist. When we say the word *set* we mean a collection of objects. But we also mean a collection that does not contain itself as a member. This is the approach we take in this book. For any set S, $S \in S$ *is always false*, that is, $S \notin S$ *is always true*. (Remember that $S \subset S$ is always true. A set is always a subset of itself.)

As for the barber/village version of the paradox, we just claim that such barbers and such villages do not exist. Not in the mathematical world anyway. Such a village, like *Brigadoon*, is a real-world notion.

BOOLEAN ALGEBRA

We have seen how simple logical statements may be combined by means of connectives, negation, implication, and other logical operations to form more complicated statements. For example, the simple statements p and q may be combined to form: $p \wedge q \Rightarrow p \vee \sim q$. Similarly, one can define operations that allow sets to be combined to form other sets. Once this is done, the resulting laws of set theoretic combination can be collected to form rules that yield an algebra for set manipulation. The collection of these basic laws is sometimes known as *Boolean algebra*, in honor of the British mathematician George Boole (1815–1864), who was one of the first to look at set theory from this point of view. In this section we will examine only the most fundamental of these operations and the resulting laws.

DEFINITION 6. Let S and T be sets. The union of S and T, written $S \cup T$, is the set of all elements that belong to S or that belong to T. The intersection of S and T, written $S \cap T$, is the collection of all elements that belong to S and that belong to T.

Therefore

$$S \cup T = \{x \colon x \in S \text{ or } x \in T\}$$

and

$$S \cap T = \{x : x \in S \text{ and } x \in T\}.$$

Definition 6, of course, turns on the key words *or* and *and*. The use of *or* in the definition of $S \cup T$ tells us the $S \cup T$ is a larger set than either S or T in the sense that S and T are both subsets of $S \cup T$. Similarly, $S \cap T$ is smaller than either S or T since it is a subset of each. Precisely, these statements compose:

THEOREM 1. (a) $S \subset S \cup T$ and $T \subset S \cup T$.
(b) $S \cap T \subset S$ and $S \cap T \subset T$.

PROOF. (In order to prove that $A \subset B$ we must show—according to definition 1—that $x \in A \Rightarrow x \in B$.)

(1) Let $x \in S$. Then $x \in S$ or $x \in T$. Thus $x \in S \cup T$. Hence $S \subset S \cup T$. Similarly, $T \subset S \cup T$.

(2) Let $x \in S \cap T$. Then $x \in S$ and $x \in T$. Thus $S \cap T \subset S$ and $S \cap T \subset T$. This completes the proof.

Here is a simple example. Let $S = \{2, 5, 7, 10\}$ and $T = \{2, 7, 11\}$. Then, $S \cup T = \{2, 5, 7, 10, 11\}$ and $S \cap T = \{2, 7\}$.

These operations also apply to real-world sets. Take, for example,

A = "set of all students at Lehigh who study calculus"

and

B = "set of all women at Lehigh."

Then

$A \cup B$ = "set of all students at Lehigh who study calculus or students at Lehigh who are women"

and

$A \cap B$ = "set of all women at Lehigh who study calculus."

In any particular situation involving sets, we usually think of each given set as being a subset of some *universal set U*. The context ordinarily determines the nature of U. For example, suppose we were given sets S, T, V, . . . , each of which contains some of the natural numbers. Then it would be appropriate to think of these sets as belonging to a universe consisting of the set $U = \mathbb{N}$. This interpretation allows us to make sketches of the given sets as portions of a rectangle whose name is U. Whenever a general picture of this type is drawn, the sets are often indicated as the interior of circular regions and the resulting sketch is called a *Venn diagram*. A Venn diagram with two sets S and T is given in figure 21. $S \cap T$ is shown as the double-shaded region and the larger set $S \cup T$ is given by the diagonal shading.

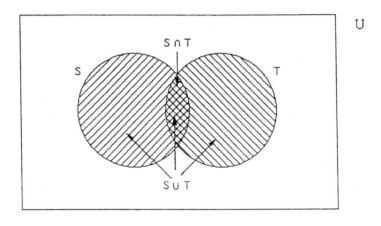

Figure 21: $S \cup T$ and $S \cap T$

Frequently, we are interested in the collection of elements that do not belong to a particular set S. The operation involved here is called *complementation* and the resulting set is called the *complement* of the given set. In fact, there are two closely related notions of the complement of a set.

DEFINITION 7. Let S and T be subsets of some universal set U.
(1) The complement of T with respect to S (or the complement of T in S), denoted by $S \backslash T$, is defined by

$$S \backslash T = \{x: x \notin T \text{ and } x \in S\}.$$

(2) The complement of T, denoted by $\sim T$, is defined by

$$\sim T = \{x: x \notin T \text{ and } x \in U\}.$$

Notice that $\sim T = U \backslash T$.

The symbol $\backslash$ in $S \backslash T$ intentionally mimics the traditional "minus sign" so that $S \backslash T$ reminds you of "S minus T." The "wiggle sign" in $\sim T$ is identical with the logical sign for negation. Think of "$\sim T$" as "elements not in T" (but, of course, belonging to whatever the universal set may be).

Venn diagrams for $S \backslash T$ and for $\sim T$ are shown, respectively, in figures 22 and 23.

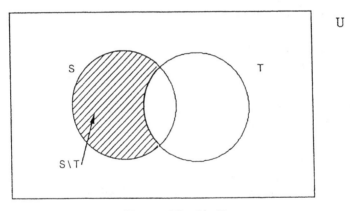

Figure 22: $S \backslash T$

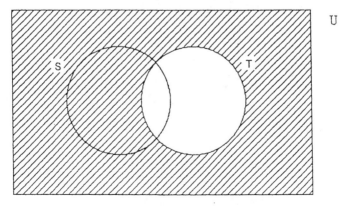

Figure 23: $\sim T = U \backslash T$

If we go to our specific examples $S = \{2, 5, 7, 10\}$, $T = \{2, 7, 11\}$, with the universal set being $U = \mathbb{N}$, we have

$$S \setminus T = \{5, 10\}$$

and

$$\sim T = U \setminus T = \{x: x \in \mathbb{N}, x \neq 2, 7, 11\}.$$

(That is, $\sim T$ is the "set of all natural numbers other than 2, 7, or 11.")

Many of the basic laws that constitute Boolean algebra flow easily from the definitions of the three fundamental operations. A few of these are given in:

THEOREM 2. For any sets S and T and a universal set U.
(i) $S \cup S = S$,
(ii) $S \cap S = S$,
(iii) $S \cup \emptyset = S$,
(iv) $S \cap \emptyset = \emptyset$,
(v) $S \cup T = T \cup S$,
(vi) $S \cap T = T \cap S$,
(vii) $\sim (\sim A) = A$,
(viii) $A \cup (\sim A) = U$,
(ix) $A \cap (\sim A) = \emptyset$,
(x) $A \cap U = A$.

PROOF (OF (i), (iv), AND (vii)):

(i)

$$(i)\ S \cup S = \{x: x \in S \text{ or } x \in S\}$$
$$= \{x: x \in S\}$$
$$= S.$$

(iv)

$$(iv)\ S \cap \emptyset = \{x: x \in S \text{ and } x \in \emptyset\}$$
$$= \emptyset, \text{ since there exists no } x \text{ with } x \in \emptyset.$$

(vii)

$$\sim A = \{x : x \notin A\}, \text{ so}$$
$$\sim(\sim A) = \{x : x \notin (\sim A)\}$$
$$= \{x : x \in A\},$$

since

$$x \notin (\sim A) \Leftrightarrow \text{``}x \text{ does not belong to } \sim A\text{''}$$
$$\Leftrightarrow \text{``}x \text{ does not belong to the set of}$$
elements that are not in A''
$$\Leftrightarrow x \in A.$$

Then, $\sim (\sim A) = A$.

The proofs of the other parts of theorem 2 are left as exercises. You can verify the correctness of each part by drawing an appropriate Venn diagram.

Other Boolean algebra laws involve more than two sets. Some of these are given in:

THEOREM 3. Let A, B, C be sets. Then
(1) $A \cup (B \cup C) = (A \cup B) \cup C$,
(2) $A \cap (B \cap C) = (A \cap B) \cap C$,
(3) $A \cup (B \cap C) = (A \cup B) \cap (A \cup C)$,
(4) $A \cap (B \cup C) = (A \cap B) \cup (A \cap C)$.

PROOF (OF PART (4)).

$$x \in [A \cap (B \cup C)] \Leftrightarrow x \in A \text{ and } x \in (B \cup C)$$
$$\Leftrightarrow x \in A \text{ and } (x \in B \text{ or } x \in C)$$
$$\Leftrightarrow (x \in A \text{ and } x \in B) \text{ or } (x \in A \text{ and } x \in C)$$
$$\Leftrightarrow x \in A \cup B \text{ or } x \in A \cap C$$
$$\Leftrightarrow x \in [(A \cap B) \cup (A \cap C)].$$

Thus,

$$A \cap (B \cup C) = (A \cap B) \cup (A \cap C).$$

The proofs of the other parts of theorem 3 are left as exercises.

Part (1) of theorem 3 says that, if you want to form the union of three sets, the order of formation does not matter. This property of the union operator is called the *associative property*. Part (2) says that $\cap$ is also an associative operator. Part (3) states that $\cup$ is *distributive* with respect to $\cap$ and part (4) says that $\cap$ is distributive with respect to $\cup$. You notice the similarity of behavior of $\cup$ and $\cap$ with that of ordinary addition and multiplication, respectively. This similarity of behavior will become clear when we append the operations of addition and multiplication to the set $\mathbb{N}$ of natural numbers.

Before we leave Boolean algebra, we must present two important results known as De Morgan's laws, after the famous British mathematician Augustus De Morgan (1806–1871) who produced a version of them around 1847.

THEOREM 4 (De Morgan's laws). Let A and B be two sets. Then

(i) $\sim (A \cup B) = (\sim A) \cap (\sim B)$

and

(ii) $\sim (A \cap B) = (\sim A) \cup (\sim B)$.

PROOF (OF PART (I)).

$$x \in [\sim(A \cup B)] \Leftrightarrow x \notin (A \cup B)$$

$$\Leftrightarrow x \notin A \text{ and } x \notin B \ (*)$$

$$\Leftrightarrow x \in (\sim A) \text{ and } x \in (\sim B)$$

$$\Leftrightarrow x \in [(\sim A) \cap \sim (B)].$$

Then, $\sim (A \cup B) = (\sim A) \cap (\sim B)$. (The step in the proof indicated by the asterisk comes from the fact that $x \in (A \cup B)$ holds if and only if $x \in A$ or $x \in B$. Thus, $x \notin (A \cup B)$ holds if and only if $x \notin A$ and $x \notin B$.)

The proof of (ii) is similar and is left as an exercise.

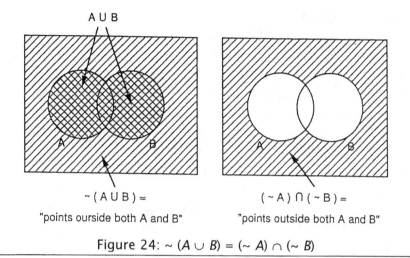

AUB

~(A U B) =

"points ourside both A and B"

(~A) ∩ (~B) =

"points outside both A and B"

Figure 24: ~ (A ∪ B) = (~ A) ∩ (~ B)

All of the results of theorems 2, 3, and 4 can be verified (not proved) by means of Venn diagrams. The first De Morgan law appears in the Venn diagrams of figure 24. The first sketch in the figure shows $A \cup B$ as a double-shaded region and then ~ $(A \cup B)$ as the diagonally shaded region. The second sketch shows, as the shaded region, the set of points that are outside A and that are outside B. But this is just $(\sim A) \cap (\sim B)$. The verification of (i) consists of the observation that the diagonally shaded regions in the two sketches of figure 24 are identical.

De Morgan's laws can be remembered by noticing that (i) and (ii) say, respectively,

"The complement of the union is the intersection of the complements"

and

"The complement of the intersection is the union of the complements."

Sounds like something Danny Kaye might have said in *The Court Jester*, but it sticks in your mind.

LOGIC AND SETS

You may have noticed that the proofs of the Boolean algebra results of theorems 2, 3, and 4 consist mainly of translating the language of the set theoretic operations into the appropriate logical language, through which the operations are defined, and then, at the end, translating the notions back into the language of sets. For example, in order to prove that $A \cap (\sim A) = \emptyset$ (theorem 2, part (ix)), we proceed as follows: $x \in A \cap (\sim A)$ if and only if $x \in A$ and $x \in (\sim A)$. But this is valid if and only if $x \in A$ and $x \notin A$. But there are no elements in the mathematical world with this property. Hence, $A \cap (\sim A) = \emptyset$. Thus, the proof involves only the proper interpretation of the logical operators *and* and *not*.

Consequently, it is natural to suspect that *there exists a fundamental connection between symbolic logic and the theory of sets.* Such a connection, in fact, exists and at the deepest levels (or highest levels, depending on your point of view) the subjects merge and become essentially indistinguishable. For our purposes, we need not explore the formal connections between the two subjects. We require only an informal understanding of the single method of associating a set P with a logical statement p. The association goes like this:

Let p be a logical statement and let U denote the set of all logical possibilities for p. Then the subset P of U that contains all possibilities for which p is true is called the *truth set* for p. As an example, let p denote the statement:

n is an even natural number.

Here, the basic idea is that the symbol n denotes a variable, that is, n represents a natural number whose value is unknown. For some values of n, statement p will be true. For other values, statement p is false. For example, p is true if $n = 16$ and false if $n = 9$. The statement tells us that $n \in \mathbb{N}$ so that the set of all logical possibilities for statement p is just the set $\mathbb{N}$. Obviously, p will be true if and only if n belongs to the set $E = \{2, 4, 6, \ldots\}$. Thus the truth set P for statement p is E.

In this elementary example, p is a simple logical statement. But the truth set notion remains valid for compound statements, that is, for logical statements composed of simple logical statements strung together by certain logical connectives and operations. As an example, consider the implication $p \Rightarrow q$.

Here, p and q are considered to be *given, but unknown*, statements. But they are statements about something or other and, for certain of these "some-things," p (respectively q) will be true and for others it will be false. As before, let U denote the set of all logical possibilities for p and q and let P and Q denote, respectively, the truth sets for p and q. What then is the truth set for $p \Rightarrow q$?

We know (see figure 2) that $p \Rightarrow q$ is true except in the single case where p is true and q is false. Hence, the truth set for $p \Rightarrow q$ will consist of all the elements of U except for those which belong to P and not to Q, that is, except for those elements that belong to $P \cap (\sim Q)$. So, if we denote the truth set of $p \Rightarrow q$ by T, then $T = U \backslash [P \cap (\sim Q)]$. This is the same as $\sim [P \cap (\sim Q)]$, which by one of De Morgan's laws (theorem 4) is $\sim P \cup (\sim(\sim Q)) = (\sim P) \cup Q$. Thus $T = \sim P \cup Q$. Hence, the truth set for $p \Rightarrow q$ consists of those elements that are outside P or inside Q. This truth set is shown in the Venn diagram of figure 25.

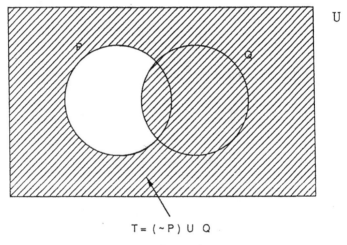

$$T = (\sim P) \cup Q$$

Figure 25: Truth set for $p \Rightarrow q$

Admittedly, this discussion has been brief and abstract. I do not want to press the issue too far. The notion you should grasp is that of associating a set—the *truth set*—with a particular logical statement. At this time, you need not struggle with particulars. Later on—whenever we need these ideas in any particular case—we will develop the truth set from first principles.

For now, we have lingered long enough over sets and logic. We must get on to mathematical analysis. This means we need to turn our attention to the material from which mathematics is made. Time now to look at numbers.

Chapter 3

BEYOND COUNTING

—Milton wondered how the stars dance:

What if the sun
Be Centre to the World, and other Stars
By his attractive virtue and their own
Incited, dance about him various rounds.[1]

Something stirred the stars in John Milton's day and something stirs them now. He wondered what moved them and most of us—even with Newton and Einstein and big bang cosmology at hand—wonder still. But the whirling of real-world stars is not our present concern. Our eyes are on other quarry. We seek mathematical truth. We want to know what constitutes the fixed-point center of the *mathematical world*. Around what do the mathematical stars turn? How do equations dance? What spark excites mathematical evolution?

Milton could only wonder about the center of his world. We need not. The deep heart's core of the mathematical world is well known. It is denoted by the symbol $\mathbb{N}$ and its name is "the set of natural numbers." Around the natural numbers the mathematical world turns like a wheel.

Children instinctively point out the elements of $\mathbb{N}$ when they count. And, so far, we have done little more than they. But pointing is not mathematics and to proceed we need something sharper than instinct. We must set down a precise description of the natural numbers. Fortunately, someone already has done this for us. His name is Guiseppe Peano (1858–1932) and in 1889 he produced five axioms that completely describe the natural numbers.

THE PEANO AXIOMS. The set $\mathbb{N}$ of natural numbers has the following properties:

(A) $\mathbb{N} \neq \emptyset$, it contains an object called 1.

81

(B) For each $x \in \mathbb{N}$, there exists a unique $x' \in \mathbb{N}$, called the successor of x.

(C) $x' \neq 1$ for any $x \in \mathbb{N}$.

(D) $x' = y' \Rightarrow x = y$ for any $x, y, \in \mathbb{N}$.

> (Traditionally, the symbol x' is read "x prime." This is merely a manner of speaking and we will soon come to an important, and nonrelated, use of the word "prime." Similarly, y' is read "y prime." The general symbol $x \neq y$ is called an "inequality" and means "x is not equal to y.")

(E) Let $S \neq \varnothing$ and $S \subset \mathbb{N}$. Suppose

> (a) $1 \in S$, and
> (b) $x \in S \Rightarrow x' \in S$

Then

$$S = \mathbb{N}.$$

Earlier, I spoke of "fundamental principles" from which all mathematics flows. The Peano axioms are among the most fundamental of these. These axioms tell us the *essence* of the natural numbers. In a sense, they constitute the Genesis 1 of mathematics. From the axioms—and almost from these alone—one can *deduce* (with the aid of appropriate definitions) all the properties of the natural numbers and then, subsequently, the properties of the richer numbers that are formed from them. Exactly such a development can be found in Edmund Landau's book *Foundations of Analysis.*[2] Beginning with the Peano axioms, Professor Landau takes the reader—in only 134 pages—successively from the natural numbers to the integers to the rational numbers and to the real numbers and, finally, to the complex numbers. Landau masterfully presents Peano's program for the development of the number system in such a manner as to require almost no mathematical background. In the preface, the author asserts that the reader need no technical background, only the ability to think logically.

However, Landau's presentation—and, indeed any rigorous development of Peano's program—requires of the reader considerable mathematical sophistication. (Mathematical sophistication is independent of technical background. Background refers to how much mathematics you know. Sophistication refers to how deeply you know it.) Peano's method lies deep and the beginner feels the pressure. In my view, beginners should come to

Landau only after they have seriously explored mathematics in shallower waters. You would not start a budding English major's literary education with "The Cantos" of Erza Pound. For analogous reasons, we will not pursue the Peano/Landau development of the number system.

What we will do is indicate, without undue worry over rigor or detail, how such a development might go and state the basic properties of the emerging numbers. Then we will examine these properties and see what we can deduce from them.

First of all, the Peano axioms tell us that the set of natural numbers is not empty. This means that $\mathbb{N}$ contains an element. This element is assigned the name "1." I have resisted the temptation to say "$\mathbb{N}$ contains at least one element." The objective of the Peano program is to deduce the properties of the natural numbers from the axioms. At the stage when he writes: "$\mathbb{N} \neq \emptyset$," it makes no sense to speak of "one." And even after he has written "an object called 1," it is not clear what we would mean by the phrase "at least one" since there has yet been no talk about "order," that is, about some natural numbers being larger or smaller than others. For these reasons, the axioms are labeled A–E rather than 1–5. If all you have are the axioms of Peano, you do not know what the symbols 2, 3, 4, 5 mean.

Axiom B says that each natural number has a unique natural number associated with it called the "successor." So, in particular the number 1 given by axiom A has a successor $1'$, whose name we do not yet know. But axiom C says that 1 is itself not the successor of any natural number. Thus $x' = 1$ is false for all x. The contrapositive of the implication in axiom D (which is equivalent to axiom D) states $x \neq y \Rightarrow x' \neq y'$. Hence, different natural numbers have different successors.

Axiom E is called the *axiom of mathematical induction*. I will soon write this axiom in a different, but equivalent, form and we will find many uses for it in our study of mathematics. In either form, the axiom gives sufficient conditions on a set S, which ensure that it becomes the entire set of natural numbers. The conditions here are that S contains the number 1 and that, whenever S contains any number, it contains the successor of that number. (Here is a physical analog of the induction axiom: consider an infinite row of dominos standing on edge. Suppose they are placed so that, whenever any particular domino falls, the one after it falls. Now push the first domino over. Axiom D says that *all* the dominos fall.)

After the axioms, the next stage in the Peano program consists of the introduction of two operations on the set of natural numbers and the defining of the concept of order. Associated with the order concept are the familiar phrases "greater than" and "less than." The two operations are the also familiar—but perhaps vaguely understood—school-day notions of addition and multiplication.

ADDITION, MULTIPLICATION, ORDER

Addition comes from the Peano axioms by way of the notion of successor. First, we *define* the limited concept of $n + 1$, where n is any natural number. (I have changed notation for two reasons: (i) n and m are more commonly used as symbols for natural numbers than are x and y, and (ii) one of the keys to learning mathematics consists in becoming comfortable with changes of notation. Saying "let $x \in \mathbb{N}$" or "let $n \in \mathbb{N}$" makes no mathematical difference. But the latter phrase occurs more commonly.) The definition is

$$(1) \qquad\qquad n + 1 = n',$$

Thus $n + 1$ is *defined* to be the natural number that is the successor to n, whose existence is given by axiom B. This tells us, in particular, that

$$1 + 1 = 1'.$$

We know by axiom C that $1' \neq 1$. So $1'$ is a natural number different from 1. We next assign this natural number a name. The name is, of course, 2. So, *by definition*, $1' = 2$ and thus

$$1 + 1 = 2.$$

We may now apply definition (1) to the new natural number 2. This gives

$$2 + 1 = 2'.$$

We would like to give a name to the natural number. But we face a complication. We know by axiom C that $2' \neq 1$. But we must face the possibility that the successor to 2 could be 2 itself. It turns out this cannot happen. One can *prove* directly from the Peano axioms a theorem that says that $n' \neq n$ for any $n \in \mathbb{N}$. Thus the successor of a natural number is always different from the number itself. (See Landau's book for details.)

In particular, $2' \neq 2$ and we are free to give $2'$ a name different from 1 or 2. Let's call it 3. Hence,

$$2 + 1 = 3.$$

We continue in the same manner to introduce the new numbers 3, 4, 5, . . . by means of the equations:

$$2 + 1 = 2' = 3,$$

$$3 + 1 = 3' = 4,$$

$$4 + 1 = 4' = 5,$$

. . . .

It remains to define the concept of addition of arbitrary natural numbers n and m in such a manner that it is consistent with the special case of $n + 1 = n'$. We will omit the (somewhat tedious) details and simply assume that this has been done. Moreover, we will accept the fact that once the definition has been properly written, the following theorem may be proved:

THEOREM 1. If n, m, and r are arbitrary natural numbers then
(i) $m + n = n + m$ and
(ii) $n + (m + r) = (n + m) + r$.

These two well-known properties of addition have names. The statement in (i) says that addition of natural numbers is *commutative* and (ii) expresses the property that addition is *associative*. Equations (i) and (ii) are, respectively, the definitions of these terms. For example:

$$3 + 4 = 4 + 3$$

and

$$2 + (3 + 4) = (2 + 3) + 4.$$

We know, of course, that both sides of the first equality are equal to 7 and that each side of the second equation reduces to the number 9. In the second equation, the parentheses tells us the sequence of operations. For example:

$$2 + (3 + 4) = 2 + 7$$
$$= 9,$$

while

$$(2 + 3) + 4 = 5 + 4$$
$$= 9.$$

(I have used the facts $4 + 3 = 7$, $2 + 7 = 9$, $2 + 3 = 5$, and $5 + 4 = 9$ by simply recalling them from memory. They have not been proved here.) In general, conclusion (ii) of theorem 1 says that if you find $m + r$ and then add this number to n, the answer obtained will be identical with that determined by first computing $n + m$ and then adding the result to r.)

For our purposes we need not pursue further the fundamental details of addition. The main point to remember is that the natural numbers are mathematical objects whose characteristics are determined by the Peano axioms. The axioms provide the concept of successor n' of a natural number n. This leads to a definition of addition of any two natural numbers n and m such that $n + m$ becomes $n + 1 = n'$ in the special case $m = 1$. We can then deduce certain theorems concerning the notion of addition—an example of which is theorem 1. We do not need to worry over the proof of theorem 1; we only need to understand that it has been proved.

We will, however, make considerable use of theorem 1 and its extensions to richer number systems. We can, in fact, use it here to establish a simple, and often talked about, result.

Schoolchildren often exaggerate the difficulty of ideas that they do not understand. I remember being told as a child that there were only ten people on earth who understood the special relativity of Albert Einstein. I also heard—from adults as well as my schoolmates—that nobody could *prove* that 2 + 2 = 4. It just *is*, they told me. I was always skeptical about the Einstein remark, but I could not settle it one way or the other. (Had they told me that only ten people exist who understand *Finnegans Wake*, I would have agreed—then and now.) But the 2 + 2 = 4 issue is easy. To settle it, we need only understand the concept of addition, the meaning of the natural numbers 1, 2, 3, 4, and have available to us the above result called theorem 1.

THEOREM 2. 2 + 2 = 4.

PROOF.

$$
\begin{aligned}
2 + 2 &= 2 + (1 + 1) && \text{(because, } 1 + 1 = 1' = 2) \\
&= (2 + 1) + 1 && \text{(by theorem 1)} \\
&= 3 + 1 && \text{(because } 2 + 1 = 2' = 3) \\
&= 4 && \text{(because } 3 + 1 = 3' = 4).
\end{aligned}
$$

These simple lines complete the proof and end the mystery forever. *Two plus two equals four* is a true statement. But not because it was written by the hand of God or because the characteristics of the real world make it so. The mathematical statement "2 + 2 = 4" holds because it is *deducible* from the Peano axioms and from other results that themselves follow from the axioms. Theorem 2 epitomizes the notion of mathematical truth. And we must seriously consider such results because there may exist truth of no other kind. Certainly there exists none other of which I am aware.

Theorem 2 is worth lingering over for another reason. It represents, in my view, one of those mathematical results that—when completely understood—leads to the understanding of other results. For example, if you are asked to prove the two statements:

(a) 9 + 1 = 10, and
(b) 6 + 4 = 10,

an understanding of theorem 2 enables you to perceive the difference between the statements and provides a procedure for dealing with each.

As for equation (a), the proper answer is: "There is nothing to prove. By definition, $9 + 1 = 9'$, the successor to 9. And the name of $9'$ is 10. Equation (b), however, requires proof. But method of proof comes easily by analogy with the proof of theorem 2. Just write:

$$
\begin{aligned}
6+4 &= 6+(3+1) \\
&= (6+3)+1 \\
&= (6+(2+1))+1 \\
&= ((6+2)+1)+1 \\
&= ((6+(1+1)+1))+1 \\
&= (((6+1)+1)+1)+1 \\
&= ((7+1)+1)+1 \\
&= (8+1)+1 \\
&= 9+1 \\
&= 10.
\end{aligned}
$$

and the proof is complete. The sixth line of the proof replaces the quantity 6 + 4 with "six plus an appropriate string of ones." The parentheses (which come from repeated applications of theorem 1) force the appropriate sequence for naming the successors of the numbers beginning with 6 and ending with 9. From this stage on the proof is quite natural, being nothing more than a precise statement of what children do when they add six to ten by counting on their fingers.

It would, of course, be tedious to repeat this process to prove 63 + 24 = 87. (It is also tedious to do this calculation by counting on your fingers.) One of the purposes of studying arithmetic in schools is to learn an algorithm for doing this, and similarly, additions in a less tedious manner.

We should notice, in passing, that addition of natural numbers is an example of what is called a *binary operation*. (A precise definition of this notion comes later.) That is, we may properly speak only of the addition of two natural numbers to produce a third. Yet, we know it is commonplace to write expressions such as "2 + 6 + 8." How do we resolve this?

The difficulty comes from the fact that—in terms of the definition of addition—the expression 2 + 6 + 8 is indeterminate. It could mean either

$$
2 + (6 + 8)
$$

or

$$(2 + 6) + 8,$$

and it must mean one or the other since addition is a binary operation. Fortunately, theorem 1 tells us that either interpretation yields the same result. Consequently, we may write $2 + 6 + 8$ without parentheses and do the calculation in either order we please.

Similarly, an expression such as $2 + 6 + 8 + 9$ has meaning in the sense that we may add the numbers in pairs with any grouping we choose. We will, by a natural extension of theorem 1, get identical results independent of the choice of pairs. Thus, if you are told that $n_1, n_2, n_3, \ldots, n_k$ are natural numbers, then you know what to make of the equation

$$m = n_1 + n_2 + n_3 + \ldots + n_k.$$

To determine m, you simply "add the k numbers on the right-hand side" by grouping them in pairs anyway you choose. In particular, line 6 of the proof of theorem 2 could be written without parentheses as

$$6 + 4 = 6 + 1 + 1 + 1 + 1.$$

Following addition, it is possible to extract from the Peano axioms a second operation that combines any two natural numbers into a third. This second operation is called *multiplication* and is denoted by a dot, or else by juxtaposition of symbols. Thus we write

$$r = n \cdot m$$

or

$$r = nm$$

to indicate that r is the natural number resulting from the multiplication of the natural numbers n and m.

Once again, we will omit the details of the definition and the fundamental results. One of the most basic of these results is

THEOREM 3. For any natural numbers n, m, and r:

(i) $nm = mn$,

(ii) $n(mr) = (nm)\, r$,

and

(iii) $n(m + r) = mn + nr$,

(iv) $n \cdot 1 = n$.

Again, we omit the proof. But we will make considerable use of theorem 3 as we proceed.

Equation (i) says that multiplication of natural numbers is *commutative* and (ii) states that multiplication is *associative*. Equation (iii) relates multiplication to addition. The property expressed in (iii) is called the *distributive* property of multiplication. More properly, (iii) states that *multiplication of natural numbers is distributive with respect to addition.*"

(These properties *define*, for the operations of addition and multiplication, the words "commutative," "associative," and "distributive." You will recall that the same three words were used earlier with respect to the set of operators of union and intersection.)

It is not difficult to see that the distributive property may be extended to expressions like:

$$n(m + r + q) = nm + nr + nq$$

or

$$n(m + r + q + s) = nm + nr + nq + ns,$$

and so on. That is, multiplication distributes itself over any finite string of sums of natural numbers. (The word *finite* is used here only for emphasis. At this stage, it makes no sense to talk about expressions like

$$n_1 + n_2 + n_3 + \ldots$$

where the addition goes on forever.)

For example, (iii) yields

$$4(3 + 2) = 4 \cdot 3 + 4 \cdot 2.$$

(Of course, we see easily that this result holds since we know from school arithmetic that each side of the equation equals 20. The point made in part (iii) of theorem 3 is that $n\,(m + r) = nm + nr$ holds for any three natural numbers n, m, and r.)

Like addition, multiplication of natural numbers is also a binary operation, that is, it associates with any pair of natural numbers a third natural number. However, part (ii) of theorem 3 allows you to write expressions such as $n \cdot m \cdot r$ since either of the two possible interpretations:

$$n \cdot m \cdot r = n \cdot (m \cdot r)$$

or

$$n \cdot m \cdot r = (n \cdot m) \cdot r$$

give the same result. Similarly, the equation

$$m = n_1, n_2 \cdot \ldots \cdot n_k$$

has meaning for natural numbers. An extension of (ii) allows one to group the right-hand terms in pairs in any manner we please to perform the operations. Moreover, part (i) of theorem 3 asserts that we may interchange terms with impunity. In particular,

$$n \cdot m \cdot r = m \cdot n \cdot r.$$

Multiplication of natural numbers is related to addition in another important way. Suppose, for example, we want to compute $2 \cdot 6$. We know the answer is 12. But we could do it this way:

$$2 \cdot 6 = 6 \cdot 2 \qquad \text{((i) of theorem 3)}$$
$$= 6 \cdot (1+1) \quad (1+1=1'=2)$$
$$= 6 \cdot 1 + 6 \cdot 1 \quad \text{((iii) of theorem 3)}$$
$$= 6 + 6 \qquad \text{((iv) of theorem 3)}$$
$$= 12 \qquad \text{(school arithmetic).}$$

The point of emphasis lies in line 4:

$$2 \cdot 6 = 6 + 6.$$

Thus, multiplying 2 by 6 turns out to be equivalent to adding two sixes. Similarly, we could justify:

$$3 \cdot 6 = 6 \cdot 3$$
$$= 6(1+1+1)$$
$$= 6 \cdot 1 + 6 \cdot 1 + 6 \cdot 1$$
$$= 6 + 6 + 6.$$

Consequently, multiplying 3 by 6 is the same as adding three sixes. In general, one can prove (although we shall not):

THEOREM 4. Let n and m be natural numbers. Then

$$n \cdot m = m + m + m + \ldots + m$$

and

$$n \cdot m = n + n + n + \ldots + n$$

where the first sum contains exactly n terms and the second exactly m terms.

If you wish, therefore, you may compute $n \cdot m$ without multiplying at all. You simply add m to itself n times, or else you add m to itself n times.

Either way you obtain $n \cdot m$. In this sense, multiplication and addition of natural numbers depend on one another even though they often act as if they do not.

We have already used part (iv) of theorem 3 in the above when we noticed that $6 \cdot 1 = 6$, but it deserves another look. The property expressed in the statement:

$$n \cdot 1 = n, \text{ for each } n \in \mathbb{N}$$

together with

$$n \cdot 1 = 1 \cdot n$$

has a name. Because $n \cdot 1 = 1 \cdot n = n$ holds for each natural number n, we say that 1 *is the multiplicative identity* in $\mathbb{N}$. (This defines the phrase *multiplicative identity in* $\mathbb{N}$.) It would be nice if there were also an additive identity in $\mathbb{N}$, that is, if there were a natural number z with the property that $z + n = n + z = n$ for each natural number n. But, alas, no such z exists. And it is partly because of this defect that the natural numbers are insufficient for even the most elementary mathematics. Shortly, we will remedy the situation by creating a richer number system. But, before we do, we need to look at one more property of $\mathbb{N}$. We must examine what it means for one natural number to be larger or smaller than another.

If n and m are natural numbers, we say that m *is larger than* n and we write $n < m$ provided there exists a natural number r such that $m = n + r$. Hence, $5 < 8$ because $8 = 5 + 3$. Also, for any $n \in \mathbb{N}$, $n < n'$ because $n' = n + 1$.

When $n < m$, we may also write $m > n$ and say "n is less than m" or "m is greater than n." If we know that $n < m$ or $n = m$, we may write $n \leq m$. This is read "n is less than or equal to m." Equivalently, we could write $m \geq n$ and say "m is greater than or equal to n." Therefore, $5 \leq 8$ is a correct statement. So is $5 < 8$. The latter statement is preferred because it is a sharper result.

This notion introduces an *ordering* of the natural numbers that agrees with the intuitive notion of one number being larger than another. Two useful properties of the ordering are given in

THEOREM 5. Let n, m, and r be natural numbers. Then

(i) $n < m$ and $m < r \Rightarrow n < r$

and

(ii) $n < m \Rightarrow n + r < m + r$.

PROOF. (i) Suppose $n < m$ and $m < r$. Then $m = n + q$ and $r = m + s$ for some natural numbers q and s. Thus,

$$r = m + s$$
$$= (n + q) + s$$
$$= n + (q + s)$$
$$= n + d$$

where d is the natural number $d = q + s$. Hence, $n < r$.

(ii) Suppose $n < m$. Then $m = n + t$ for some $t \in \mathbb{N}$. Hence

$$m + r = (n + t) + r$$
$$= n + (t + r)$$
$$= n + (r + t)$$
$$= (n + r) + t.$$

Thus, $n + r < m + r$.

The property of $<$ expressed in (i) is called the *transitive property*. (This simply says: if n is less than m and m is less than r, then n is less than r.) For short, we may simply say: "$<$ is transitive." Part (ii) says that the relation $<$ is *preserved by addition*: Thus you may add the same number to both sides without destroying the sense of an inequality.

You should see that theorem 5 remains valid if the inequality $<$ is everywhere replaced by the weaker inequality $\leq$. (Recall that $m \leq n$ means $m < n$ or $m = n$.) Thus,

$$n \leq m \Rightarrow n + r \leq m + r.$$

and

$$n \leq m \ \Rightarrow n + r \leq m + r.$$

Before we move on, let's notice that there exists no largest natural number.

THEOREM 6. $\mathbb{N}$ contains no largest element.

PROOF. Let $n \in \mathbb{N}$ (thus n is an *arbitrary* natural number).
   Then $n' \in \mathbb{N}$ (second Peano axiom).
   But $n' = n + 1$ (definition of addition).
   So $(n + 1) \in \mathbb{N}$ and $n < n + 1$ (definition of $<$).
   Hence n is not the largest element of $\mathbb{N}$.
   But n is arbitrary, so $\mathbb{N}$ has no largest element.

This little proof merits attention for two reasons: (i) it contains references to the mathematically basic notions of *axiom*, *definition*, and *arbitrariness* and (ii) it provides a quick way to see that the set of natural numbers cannot be finite—they go on and on. (We have, in fact, anticipated theorem 6 earlier in our "proof" that $\mathbb{N}$ is an infinite set.)

THE INTEGERS

We have already noticed that the natural numbers contain no element that can serve as an "additive identity," that is, there is no $z \in \mathbb{N}$ such that $n + z = n$ for each $n \in \mathbb{N}$. (For example, there is no natural number z such that $6 + z = 6$.) To remedy this defect, we adjoin to the natural numbers a new number that, by definition, possesses this property. We denote this new number by 0 and call it *zero*. So, now we have *extended* the set of natural numbers and formed a larger set (for which there is no commonly accepted symbol) $\{0, 1, 2, 3, \ldots\}$.

But we still cannot perform certain simple algebraic operations. For example, we cannot solve the equation $x + 6 = 5$ in the new system, that is, there exists no $x \in \{0, 1, 2, 3, \ldots\}$ such that $x + 6 = 5$. (The trouble lies in the fact that $x + 6 = 6$ in case $x = 0$, and for all other x in $\{1, 2, 3 \ldots\}$, $x + 6 > x$.) Peano's program provides a way in which the defect can be remedied. What you do (see Landau for details) is adjoin to the collection $\{0, 1, 2, 3, \ldots\}$ new numbers $-1, -2, -3, \ldots$, which are called *negative numbers*.

Next, through appropriate definitions (which we shall omit), the notion of addition may be extended from $\mathbb{N}$ to this new set in such a way that, for each $n \in \mathbb{N}$,

$$n + (-n) = (-n) + n = 0.$$

Moreover, the binary operation of multiplication and the order notion $<$ (also the weaker notion $\leq$) may be extended to the new set. We are then able to form the sum $n + m$ and the product $n \cdot m$ for any two elements in the new set, and each operation yields a number that also belongs to the set. The relation $<$ orders the new set as

$$\ldots -3 < -2 < -1 < 0 < 1 < 2 < 3 \ldots.$$

When all this is done we have a set

$$\mathbb{Z} = \{\ldots, -3, -2, -1, 0, 1, 2, 3, \ldots\}$$

on which are appropriately defined the operations $+$, $\cdot$, and the order relation $<$. The set $\mathbb{Z}$ is known as the *set of integers*, or simply as the *integers*. (An integer is a member of $\mathbb{Z}$.)

The number $-n$ is called the *negative of n*, or more simply as *negative n*. (Often, and somewhat improperly, $-n$ is also called *minus n*.) Thus -6 is the negative of 6. $\mathbb{Z}$, then, is the set composed of the natural numbers, zero, and the negatives of the natural numbers.

The natural numbers now form a proper subset of the integers. In terms of the order relation, $\mathbb{N}$ consists of all those members of $\mathbb{Z}$ that are greater than zero. In this context, $\mathbb{N}$ is often referred to as the *positive integers* and is often denoted by $\mathbb{Z}^+$. Thus

$$\mathbb{N} = \mathbb{Z}^+ = \{1, 2, 3, \ldots\}.$$

The key properties of the set of integers are described in the following long theorem whose proof is omitted. (One might reasonably begin with the set of statements in the theorem and take them as *postulates* or defining characteristics for the integers. But it is better, I think, to get at least a glimpse—as we have done—of the development from things more fundamental, that is, from the natural numbers and the Peano axioms.) The definitions of certain terms such as *closed, commutative, associative,* and so on are given in parentheses alongside several of the theorem's conclusions.

(For example, A(i) ensures that the sum of any two integers is itself an integer. A mathematician would simply say: "$\mathbb{Z}$ is closed with respect to addition.")

THEOREM 7. *Let* $\mathbb{Z} = \{\ldots, -3, -2, -1, 0, 1, 2, 3, \ldots\}$ *be the integers.*

A	(i)	$m, n \in \mathbb{Z} \Rightarrow (m + n) \in \mathbb{Z}$	($\mathbb{Z}$ is *closed* with respect to addition)
	(ii)	$m, n \in \mathbb{Z} \Rightarrow m + n = n + m$	(addition of integers is *commutative*)
	(iii)	$m, n, r \in \mathbb{Z} \Rightarrow m + (n + r)$ $= (m + n) + r$	(addition of integers is *associative*)
	(iv)	$n \in \mathbb{Z} \Rightarrow n + 0 = n = 0 + n$	(0 is an *additive identity* for $\mathbb{Z}$)
	(v)	If $n \in \mathbb{Z}$ then there exists $u \in \mathbb{Z}$, such that $n + u = 0$ $= u + n$	(u is an *additive inverse for n.* We write $u = -n$.)
M	(i)	$m, n \in \mathbb{Z} \Rightarrow m \cdot n \in \mathbb{Z}$	($\mathbb{Z}$ is *closed* with respect to multiplication)
	(ii)	$m, n \in \mathbb{Z} \Rightarrow m \cdot n = n \cdot m$	(multiplication of integers is *commutative*)
	(iii)	$m, n, r \in \mathbb{Z} \Rightarrow m \cdot (n \cdot r)$ $= (m \cdot n) \cdot r$	(multiplication of integers is *associative*)
	(iv)	$n \in \mathbb{Z} \Rightarrow n \cdot 1 = n = 1 \cdot n$	(1 is a *multiplicative identity* for $\mathbb{Z}$)
D	(i)	$m, n, r \in \mathbb{Z} \Rightarrow m (n + r)$ $= mn + mr$	(multiplication is *distributive with respect to addition*)
O	(i)	If $m, n, r \in \mathbb{Z}$ then $m > n, n > r \Rightarrow m > r$	(> is *transitive*)
	(ii)	If $m, n, r \in \mathbb{Z}$ then $m > n \Rightarrow m + r > n + r$	(> is *preserved by addition*)
	(iii)	If $m, n, r \in \mathbb{Z}$ and $r > 0$ then $m > n \Rightarrow mr > nr$	(> is *preserved by multiplication by positive integers*)
	(iv)	If $m, n, r \in \mathbb{Z}$ and $r > 0$ then $mr > nr \Rightarrow m > n$	(A *common positive factor may be cancelled from an inequality*)

Admittedly, theorem 7 contains lots of information. But much of this consist of statements that are formally the same as analogous statements for $\mathbb{N}$. Moreover, the theorem will be restated in an equally analogous form when we come to the number systems that are richer than the integers. You will have, consequently, ample opportunity to become familiar with the theorem's conclusions and with the terms defined parenthetically with them. The theorem is broken into four parts labeled A, M, D, and O, respectively. The labels stand respectively for "addition," "multiplication," "distributivity," and "order." The results in A describe properties of addition, those of M give properties of multiplication. The single result of D is the so-called *distributive* property of multiplication that relates the operations of addition and multiplication. The statements in O describe fundamental properties of the order relation and its relation with addition and multiplication. We shall make no attempt to prove any part of theorem 7. We need only the results of the theorem and the understanding that the theorem can be proved (actually in an elementary manner) and each step of the proof can be traced back step-by-step all the way to the Peano axioms. These properties of the integers come to us neither through mystagogy nor higher authority. They hold because they can be proved.

Even though we will not prove any part of theorem 7, occasionally we will want to prove some simple consequences of the theorem. An example is:

THEOREM 8. Let $n, m, r \in \mathbb{Z}$. Then

$$n + r = m + r \Rightarrow n = m.$$

PROOF. $n + r = m + r \Rightarrow$

$$(n + r) + (-r) = (m + r) + (-r)$$

$$\Rightarrow$$

$$n + (r + (-r)) = m + (r + (-r)) \qquad \text{(valid by Theorem 7, A (iii))}$$

$$\Rightarrow$$

$$n + 0 = m + 0 \qquad \text{(valid by theorem 7, A (v))}$$

$$\Rightarrow$$

$$n = m \qquad \text{(valid by theorem 7, A (iv))}.$$

This completes the proof except for the possibly questionable first step. This step amounts to the validity of adding the same integer (here, $-r$) to both sides of an equality. This procedure clearly holds if we notice that—taking away the symbols peculiar to theorem 8—it amounts only to the implication: $u = w \Rightarrow u + v = w + v$, which holds by definition of addition.

Theorem 8 represents a type of *cancellation law:* if $n + r = m + r$, then the r can be cancelled from both sides to yield $n = m$. Such a law has many applications. One of the first—for us—involves the removal of the word "and" in the parenthetic remark of part A (iv) of theorem 7.

We know—by the very definition of the set of integers—that zero plays the role claimed for it in A (iv). Thus, 0 is an identity for $\mathbb{Z}$. (This point of view, of course, makes the proof of A (iv) trivial). But what about the word "and" in the parenthetic remark? Can there exist another element of $\mathbb{Z}$, say, w, such that $n + w = n$ for each integer n? In other words, is the identity in $\mathbb{Z}$ *unique?*

THEOREM 9. The additive identity in $\mathbb{Z}$ is unique.

PROOF.

By hypothesis,	$n + w = n$.
Also	$n + 0 = n$.
Hence	$n + w = n + 0$.
So	$w + n = 0 + n$.
Theorem 8 gives	$w = 0$.

This completes the easy proof and we may now, if we wish, replace the word "and" in the parenthetic remark with "the." The integers have a unique additive identity and its name is 0.

Similarly, we can show that the additive inverse ensured by theorem 7, part A(v) is also unique.

THEOREM 10. Let $n \in \mathbb{Z}$. Suppose there exist u and v with

$$n + u = 0$$

and

$$n + v = 0.$$

Then

$$u = v.$$

PROOF. $n + u = 0$ and $n + v = 0$ give

$$n + u = n + v.$$

Theorem 8 then gives

$$u = v.$$

Please notice the difference in the two concepts of "uniqueness" expressed in theorems 9 and 10. Theorem 9 assures us that there exists only a single element z in $\mathbb{Z}$ such that $n + z = n$ for each $n \in \mathbb{Z}$. We call z "zero" and write $z = 0$. Thus, 0 is a fixed member of $\mathbb{Z}$ with the property that $n + 0 = n$ for every $n \in \mathbb{Z}$. The element 0 *does not depend on n*, that is, 0 is a *constant*. On the other hand, the additive inverse u, given by theorem 7 part A(v), *depends* on the given integer n. That is, for each n there exists an element of $\mathbb{Z}$, denoted by u, such that $n + u = 0$. Thus, the additive inverse u is a *variable* that varies as n varies. We see this dependence explicitly in the notation $u = -n$. Theorem 10 tells us that the additive inverse is unique in the sense that, each particular $n \in \mathbb{Z}$ possesses only one additive inverse. That is, if v has the property that $n + v = 0$, then $v = -n$.

Another point needs to be made about theorem 7, part A (v). The parenthetic remark says that the additive inverse of any integer n will be denoted by $-n$. So, in particular, we denote the additive inverse of 1 by -1, of 2 by -2, of 3 by -3, and so forth. This represents nothing new; we introduced the negative numbers precisely for the purpose of serving as additive identities for the natural numbers. But A(v) tells us to denote the additive inverse of *any* integer n by $-n$. In particular, the additive inverse of -1 is $-(-1)$. Similarly, $-(-2)$, $-(-3)$, $-(-4)$, ... are, respectively, the additive inverses for $-2, -3, -4, \ldots$.

This all seems strange at first glance. After all, the list

$$\ldots, -3, -2, -1, 0, 1, 2, 3, \ldots$$

composes the entire collection of integers, provided we properly interpret the three dots on each side. Nowhere in this list does there appear a symbol like

– (– 1) or – (– 2). Where then do we find these numbers? According to theorem 7, part A (v), they must appear somewhere.

Not to worry: it turns out that – (– 1) = 1, – (– 2) = 2, – (– 3) = 3, and so forth. In general – (– n) = n, for each integer *n*. Moreover, the proof of this interesting fact falls within our reach. The proof is easy but subtle. Look at it carefully.

THEOREM 11. Let $n \in \mathbb{Z}$. Then $-(-n) = n$.

PROOF. There exists $u \in \mathbb{Z}$ such that

$$n + u = 0 \quad \text{(by theorem 7, A(v))}.$$

But

$u = -n$ (– n is the name of *u* by the parenthetic remark in theorem 7, A(v)).

So

$$n + (-n) = 0.$$

Hence

$$(-n) + n = 0 \quad \text{(commutative property of addition)}.$$

Thus *n* is the unique additive inverse of (– n) (by theorem 10).

Hence

$n = -(-n)$ (– (–n) is the name of the additive inverse of – n).

The key to the proof—and its only subtle point—lies in the last step: the transition from $(-n) + n = 0$ to $n = -(-n)$. It might be helpful to think of this in a more general setting: if *p* and *q* are integers and if $p + q = 0$, then we know that *q* is an additive inverse for *p* and we write $q = -p$ (theorem 7, A (v)). But $p + q = 0$ gives $q + p = 0$ and the second equation tells us that $p = -q$. If we now replace *p* by – n and *q* by *n*, we see that $p + q = 0$ holds since it is just $(-n) + n = 0$. Then $q = -p$ becomes $n = -(-n)$.

Notice that each integer, except for 0, has an additive inverse different from itself. Zero, however, is its own inverse since $0 + 0 = 0$. Thus, $- 0 = 0$. ($0 + 0 = 0$ holds by theorem 7, part A(iv). Then $- 0 = 0$ by A (v).)

We can now relate some of the properties of multiplication to the notion of additive inverse:

THEOREM 12. Let n be an arbitrary integer. Then

(A) $n \cdot 0 = 0$

   and

(B) $n (-1) = - n.$

PROOF. (i) $0 + 0 = 0.$

Thus
$n (0 + 0) = n \cdot 0.$

Thus
$n \cdot 0 + n \cdot 0 = n \cdot 0$ (distributive property of multiplication).

But
$n \cdot 0 = n \cdot 0 + 0.$

Thus
$n \cdot 0 + n \cdot 0 = n \cdot 0 + 0.$

Therefore
$n \cdot 0 = 0$ (theorem 8).

(ii)
$n + n (-1) = n \cdot 1 + n (-1)$ (theorem 7, M (iv))
$= n (1 + (- 1))$ (distributive property)
$= n \cdot 0$ ($- 1$ is additive inverse of 1)
$= 0$ (just proved in (A)).

Then, $n + n(-1) = 0$ gives

$$n(-1) = -n \text{ (theorem 7, A (v))}.$$

Since multiplication of integers is commutative, we see also that

$$(-1)n = -n.$$

Thus, the additive inverse of n has three equivalent representations: $-n$, $(-1)n$, and $n(-1)$. For example, $-6 = 6(-1) = (-1)6$.

We have already noticed that addition and multiplication of natural numbers are related in the sense that multiplication of n by m is equivalent to adding n to itself m times (or adding m to itself n times). In particular, $3 \cdot 2 = 3 + 3$, or $3 \cdot 2 = 2 + 2 + 2$. This relationship between addition and multiplication also holds for the integers—up to a point. For example, $3(-2) = (-2) + (-2) + (-2) = -6$. But what about $(-3)(-2)$? Obviously, we cannot "add -3 to itself -2 times." Nor can we "add -2 to itself -3 times." What then can we make of an expression like $(-3)(-2)$? We know, of course, that $(-3)(-2)$ exists and is an integer. ($\mathbb{Z}$ is closed with respect to multiplication by theorem 7, part M (i).) But what integer is it?

In schools, children "learn" that $(-3)(-2) = 6$. They see this as a special case of the rule "the product of two negative numbers is always positive." The rule is simply told to them. They may doubt it only at their peril.

But telling is not teaching. And it is certainly not mathematics. Fortunately, we can do better. We have at hand sufficient machinery to settle the issue $(-3)(-2) = 6$. Here's one way:

Let

$$n = 3 \cdot 2 + 3(-2) + (-3)(-2).$$

Although the expression for n involves the "addition" of three integers and addition is a binary operation, it is not indeterminate since the associative property ensures that we obtain the same answer no matter what the sequence of additions. Now write

$$n = 3 \cdot 2 + \left[3(-2) + (-3)(-2) \right].$$

(Often we will write—for purposes of clarity—outside parentheses as square brackets or even curly braces. Here brackets are used.) Thus,

$$n = 3 \cdot 2 + (-2)[3 + (-3)]$$
$$= 3 \cdot 2 + (-2) \cdot 0$$
$$= 3 \cdot 2 + 0$$
$$= 3 \cdot 2.$$

Also

$$n = \left[3 \cdot 2 + 3(-2) \right] + (-3)(-2)$$
$$= 3 \left[2 + (-2) \right] + (-3)(-2)$$
$$= 3 \cdot 0 + (-3)(-2)$$
$$= 0 + (-3)(-2)$$
$$= (-3)(-2).$$

Since both expressions represent the number n, they must be equal. Thus

$$3 \cdot 2 = (-3)(-2).$$

Two points need to be made about this little demonstration. First, the steps involve manipulations we know to be valid, either by theorem 7, which we have assumed true, or by results we have proved from theorem 7. Second, the integers 3 and 2 here play no special role. Evidently, we could give the same demonstration beginning with arbitrary integers p and q.

THEOREM 13. Let p and q be integers. Then $(-p)(-q) = p \cdot q$.

PROOF.

$$\text{Let } n = pq + p(-q) + (-p)(-q).$$

Then

$$n = pq + [p(-q) + (-p)(-q)]$$
$$= pq + (-q)[p + (-p)]$$
$$= pq + (-q) \cdot 0$$
$$= pq + 0$$
$$= pq.$$

Also,

$$n = [pq + p(-q)] + (-p)(-q)$$
$$= p[q + (-q)] = (-p)(-q)$$
$$= p \cdot 0 + (-p)(-q)$$
$$= 0 + (-p)(-q)$$
$$= (-p)(-q).$$

Therefore,

$$pq = (-p)(-q).$$

(The justifications for the steps of the proof are analogous to those given for the special case above and are left as exercises.)

The proof of theorem 13 is easy and straightforward after the initial step. But the first sentence of the proof seems unmotivated. Where did the number $n = pq + p(-1) + (-p)(-q)$ come from. Who would know to write it down? The defect—if indeed it seems to be one—can be removed by giving a second proof that does not depend on any "rabbit out of the hat" moves.

PROOF II OF THEOREM 13.

Let p and q be integers. We know, by theorem 12 (ii) that

$$n(-1) = -n.$$

for any integer n. Thus, if $n = -1$ we have

$$(-1)(-1) = -(-1).$$

Hence, by theorem 11,

$$(-1)(-1) = 1.$$

Then, we have by theorem 12 again

$$\begin{aligned}
(-p)(-q) &= [(-1)p][(-1)q] \\
&= [(-1)(-1)][pq] \\
&= 1 \cdot pq \\
&= pq.
\end{aligned}$$

Notice that in the third line from the end of the proof, we have used both the associative and the commutative properties of multiplication. Ordinarily, a working mathematician manipulates integers without excessive use of parenthesis and without explicit justification of each step. In the future we will often be more casual with our own manipulations. We will not always justify each step in a particular calculation. We want to learn to manipulate mathematical symbols without the tedium of undue rigor. But we also want sufficient understanding so that we can provide appropriate rigor when necessary.

For example, if we are given the equation

$$4 + x + 9 = 2,$$

we will immediately assert that $x = -11$. However, if we are challenged, we can produce a proof.

1. $4 + x + 9 = 2 \Rightarrow x = -11$.

PROOF. $4 + x + 9 = 2 \Rightarrow (4 + x) + 9 = 2$

$\Rightarrow (x + 4) + 9 = 2$ (addition is commutative)

$\Rightarrow x + (4 + 9) = 2$ (addition is associative)

$\Rightarrow x + 13 = 2$

$\Rightarrow (x + 13) + (-13) = 2 + (-13)$

$\Rightarrow x + [13 + (-13)] = -11$

$\Rightarrow x + 0 = -11$

$\Rightarrow x = -11.$

The last three steps in this argument, incidentally, are easily generalized to yield the rule taught in schools called "transposition": *You may transpose an integer from one side of an equality to another if you change its sign.* Formally, this becomes

THEOREM 14. If x and a are integers then

$$x + a = 0 \Rightarrow x = -a.$$

PROOF.

$$x + a = 0 \Rightarrow (x + a) + (-a) = 0 + (-a)$$
$$\Rightarrow x + [a + (-a)] = -a$$
$$\Rightarrow x + 0 = -a$$
$$\Rightarrow x = -a.$$

Hereafter, we will not go through these formal steps. In any particular situation, we simply transpose whenever we wish. For example,

$$x + 4 = 0 \Rightarrow x = -4$$

and

$$x + (-5) = 0 \Rightarrow x = -(-5) = 5.$$

Incidentally, the equation

$$x + (-5) = 0$$

normally appears in the form

$$x - 5 = 0.$$

The second equation—expressed in terms of ordinary *subtraction*—is by definition identical to the first. Thus, in general, an equation of the form

$$x - a = b$$

means

$$x + (-a) = b,$$

which gives

$$x = a + b.$$

Notice also that, even though we are still dealing with integers, we have eased back into a notation that uses the letter "x" rather than $n, m, p, q, \ldots$ with which we have become familiar. The shift in notation occurs only for reasons of convention: traditionally x stands for an unknown. Consequently, it is natural to use x as the symbol or an integer prescribed by an equation. We then manipulate the equation properly and determine the value of x. For example, $x - 5 = 0$ yields $x = 5$. This value for x is called the *solution* of the equation $x = 5$. The simple manipulation needed to determine this value is the process of *solving* the equation. But please understand that the notation plays no role in the solution. The equation $n - 5 = 0$ is in every way equivalent to $x - 5 = 0$. The first equation has as solution $n = 5$ and the second $x = 5$; except for notation, no difference exists between them.

The integers accommodate a fair amount of arithmetic. Within $\mathbb{Z}$ we are able to perform the operations of addition and multiplication and we are assured by theorem 6 that the result of any sequence of these operations will invariably lead us back to another integer. Moreover, if we define $n - m$ by

$$n - m = n + (-m),$$

then we have essentially defined on $\mathbb{Z}$ a third operation called subtraction. Thus we can solve, within $\mathbb{Z}$, equations of the form $x + a = b$. The solution $x = b - a$ is obtained by either of two equivalent manipulations: add $-a$ to each side of $x + a = b$ or transpose the number a to the right-hand side of the equation, where it becomes $-a$. Taken together the three operations and the equation-solving capacity give a rich structure.

But not rich enough. With only the integers at our disposal we remain severely limited. For example, we cannot solve the simple equation $2x = 1$. That is, there exists no integer x such that $2x = 1$. (Verification requires only a straightforward check: $2 \cdot 0 = 0$, so $x = 0$ is not a solution. If $x \geq 1$, then

$2x \geq 2$ so no positive integer has the property that $2x = 1$. And if $x \leq -1$, then $2x \leq -2$, so no negative integer satisfies the equation $2x = 1$.) Therefore, if we want to solve equations of this type—and we do—we require more numbers. The necessary numbers are called *rational numbers*. Their development constitutes the next step in the Peano program that began with the natural numbers and led us to the integers.

THE RATIONAL NUMBERS

Here, the word "rational" means "ratio" and formally a rational number is a number represented as the ratio of one integer to another. Examples of rational numbers are 2/3 and – 5/7. In general, a rational number is a number of the form n/m, where n and m are integers and, as we shall see, we must restrict m so that $m \neq 0$. n/m is commonly called a *fraction*. The number n is the *numerator* of the fraction and m is the fraction's *denominator*. (For typesetting purposes the symbol n/m often replaces the more common symbol $\frac{n}{m}$. They mean exactly the same thing.)

The logical development of the rational numbers from the integers flows along lines similar to the construction of the integers from the natural numbers. We are not interested in the details of the construction (which can be found in the book by Landau), only in the general features and the basic results.

Essentially, the Peano process leads to the definition of a rational number as a pair of integers (n, m) where the second element of the pair must be nonzero. Operations of addition and multiplication are then defined on the set consisting of all such pairs. These operations denoted by $+$ and $\cdot$, respectively, are then examined and their basic properties determined. Almost immediately, in order to conform to traditional notation, the pair (n, m) is replaced by the fractional symbol n/m. The properties of addition and multiplication are then rewritten in the new notation. Moreover, the development process naturally leads to the requirement that *certain seemingly different fractions should determine the same rational number.* That is, two fractions n/m and p/q, say, may represent the same rational number. This occurs, it turns out, whenever $nq = mp$. This leads to the first important definition.

DEFINITION 1. Let m, n, p, q be integers with $m \neq 0$ and $q \neq 0$. Then

$$\frac{n}{m} = \frac{p}{q} \Leftrightarrow nq = mp.$$

For example,

$$\frac{2}{3} = \frac{10}{15}$$

because

$$2(15) = 30 = 3(10).$$

Notice the dual usage of the notion of equality in the above definition. The definition asserts that

$$\frac{n}{m} = \frac{p}{q} \text{ if and only if } nq = mp.$$

This seems clear enough, but we must understand that the "equal signs" on either side of the double implication have different meanings. On the left side, the symbol "=" *stands for the notion that is being defined.* On the right, "=" *denotes ordinary equality between integers.* We might appropriately invent a new symbol for the "equality" on the left-hand side and use it consistently whenever we want to indicate that two rational numbers have the same value. But to do so would violate convention and would greatly increase the complexity of our notation. Consequently, we use identical symbols for the two notions in the double implication: the old equality on the right and the equality being defined on the left.

From the outset, we have tacitly assumed that the notion of equality for the natural numbers, and thus for the integers, satisfies ordinary intuitive properties. In particular, we have supposed that for any integers p, q, r we have:

$$p = p,$$

$$p = q \Rightarrow q = p,$$

and

$$p = q, \ q = r \Rightarrow p = r.$$

These properties of equality are known, respectively, as the *symmetric property*, the *reflexive property*, and the *transitive property*. In a rigorous development of mathematics, these properties would be carefully examined (perhaps after first being given as definitions).

But we are now dealing with new objects (rational numbers) and a new notion of equality (given by definition 1), and it is natural to ask whether or not the above properties still hold. Fortunately, the answer is "yes." The proofs are easy but subtle enough to merit attention. The proof that the new equality possesses the reflexive property follows.

Let s and t be rational numbers. Then

$$s = \frac{a}{b} \text{ and } t = \frac{c}{d}$$

for some integers a, b, c, d with $b \neq 0$ and $d \neq 0$. Hence,

$$s = t \Rightarrow \frac{a}{b} = \frac{c}{d},$$

$\Rightarrow ad = bc$ (definition 1)

$\Rightarrow da = cb$ (multiplication of integers is commutative)

$\Rightarrow cb = da$ (reflexive property of integer equality)

$\Rightarrow \dfrac{c}{d} = \dfrac{a}{b}$

$\Rightarrow t = s.$

Thus, definition 1 forces (or allows, depending on your point of view) the rational numbers to inherit the reflexive property of equality from the corresponding property for ordinary equality. (Please notice that the implications in the above proof actually are double implications.) The validation of the symmetric and transitive properties of rational number equality are left as exercises.

In schools rational numbers are usually called *fractions* and children are told (often without explanation) that nonzero common factors may be cancelled from numerator and denominator. For example:

$$\frac{6}{10} = \frac{2 \cdot 3}{2 \cdot 5} = \frac{3}{5}.$$

The legitimacy of this practice follows easily from the above definition of equality. To see this, suppose a, b, c are integers with $b \neq 0$, $c \neq 0$. Then

$$\frac{a \cdot c}{b \cdot c} = \frac{a}{b}$$

because

$$(a \cdot c) \cdot b = (b \cdot c) \cdot a$$

and the second equation is precisely the *definition* of the equality in the first equation. Then the common factor $c \neq 0$ can properly be cancelled from the fraction $(a \cdot c) / (b \cdot c)$.

Notice the restrictions $b \neq 0$, $c \neq 0$ in the equations

$$\frac{a \cdot c}{b \cdot c} = \frac{a}{b}.$$

Clearly, we must have $b \neq 0$ since otherwise the symbol a/b would not represent a rational number. (By *definition*, r is a rational number if and only if $r = a/b$ with $a, b \in \mathbb{Z}$ and $b \neq 0$. We will shortly come to a reason for the requirement that the denominator of a rational number shall be nonzero.)

We must also have $c \neq 0$ since otherwise the denominator of $a \cdot c/b \cdot c$ would be zero and, hence, ac/bc would itself not be a rational number. That is $c = 0$ gives $b \cdot c = b \cdot 0 = 0$. This fact is important enough to state as:

THEOREM 15. Let $n \in \mathbb{Z}$. Then $n \cdot 0 = 0$.

PROOF.

$$n \cdot 0 = n \cdot 0 + 0$$

and

$$n \cdot 0 = n(0 + 0).$$

Then

$$n \cdot 0 + 0 = n \cdot (0 + 0)$$
$$= n \cdot 0 + n \cdot 0.$$

But we know from theorem 8 that $n \cdot 0 + 0 = n \cdot 0 + n \cdot 0 \Rightarrow 0 = n \cdot 0$.

The next stage in the Peano development of the rational numbers consists of writing appropriate definitions of addition and multiplication for pairs of these new numbers:

DEFINITION 2. Let

$$s = \frac{n}{m} \text{ and } t = \frac{p}{q}$$

be rational numbers. Then $s + t$ (addition) and $s \cdot t$ (multiplication) are defined by:

(i) $s + t = \dfrac{n \cdot q + m \cdot p}{m \cdot q}$

and

(ii) $s \cdot t = \dfrac{n \cdot p}{m \cdot q}$.

Before we proceed, let us focus on two properties of definition 2. First of all, it is a *definition*.

Nothing here requires proof. Equations (i) and (ii) describe the very meaning of addition and multiplication of rational numbers. If you want to add or multiply two rational numbers, then definition 2 tells you how. Second, the definition incorporates (as definition 1 did for equality) the dual usage of symbols for addition and multiplication. Since s and t are rational numbers, the numbers n, m, p, and q are all integers (with $m \neq 0$ and $q \neq 0$). Consequently, the "+" symbol on the right-hand side of (i) represents ordinary integer addition. The "+" sign of the left-hand side of (i), however, stands for the new operation now being defined. Similarly, the "dot" on the right-hand side of both (i) and (ii) stands for ordinary integer multiplication while the symbol "$\cdot$" on the left side of (ii) represents the new multiplication being defined by the equation.

Often, as you know, multiplication of integers is denoted by juxtaposition: $n \cdot m = nm$. The practice also holds for rational numbers although parentheses may be used for clarity. Thus (i) and (ii) may appear, respectively as:

$$\frac{n}{m} + \frac{p}{q} = \frac{nq + mp}{mq}$$

and

$$\left(\frac{n}{m}\right)\left(\frac{p}{q}\right) = \frac{np}{mq}.$$

As an example consider:

$$\frac{2}{3} + \frac{4}{9} = \frac{2 \cdot 9 + 3 \cdot 4}{3 \cdot 9}$$

$$= \frac{18 + 12}{27}$$

$$= \frac{30}{27}$$

and

$$\left(\frac{2}{3}\right)\left(\frac{4}{9}\right) = \frac{2 \cdot 4}{3 \cdot 9}$$

$$= \frac{8}{27}.$$

Recall that earlier we spoke of addition and multiplication as being *binary operations* on the set $\mathbb{N}$ of natural numbers. In particular this means that if n and m are natural numbers, then $n + m$ is a *unique* natural number. This says two things: (a) $\mathbb{N}$ is *closed* with respect to addition and (b) the sum of two natural numbers is a *uniquely determined* natural number. A similar statement holds for multiplication on $\mathbb{N}$.

Moreover, when we extended the natural numbers $\mathbb{N}$ to the integers $\mathbb{Z}$, we saw that both multiplication and addition on $\mathbb{Z}$ are binary operations. It is not difficult to *prove* from definition 2 that the addition and multiplication are also binary operations on the set of rational numbers. Ordinarily, the rational numbers are denoted by the symbol $\mathbb{Q}$. Hence

$$\mathbb{Q} = \left\{ \frac{n}{m} : n, \, m \in \mathbb{Z}, \, m \neq 0 \right\}.$$

In terms of this notation, the above discussion of binary operations says, in particular,

$$s, \, t \in \mathbb{Q} \Rightarrow (s + t) \in \mathbb{Q} \text{ and } s \cdot t \in \mathbb{Q}.$$

(Thus $\mathbb{Q}$ is *closed* with respect to addition and multiplication.)

The special rational numbers of the form $n/1$ look like integers and in a sense they are. That is, it is possible to identify each rational number $n/1$ with the integer n in such a manner that the notions of addition and multiplication of pairs of these numbers are consistent with the earlier concepts of addition and multiplication of integers. We use ordinary equality to represent this identification and we write:

$$\frac{n}{1} = n$$

for each integer n.

For example, we may think of the integer 6 as being the same as the rational number 6/1 and of 7 as being 7/1. Thus

$$\frac{6}{1} = 6 \text{ and } \frac{7}{1} = 7.$$

The comments about the consistency of addition tells us that the sum of integers

$$6 + 7 = 13$$

is equivalent to the rational number addition

$$\frac{6}{1} + \frac{7}{1} = \frac{6 \cdot 1 + 7 \cdot 1}{1 \cdot 1}$$
$$= \frac{6 + 7}{1}$$
$$= \frac{13}{1}$$

since

$$\frac{13}{1} = 13.$$

Consequently, it makes no difference whether we compute 6 + 7 using ordinary integer addition or think of them, respectively, as the rational numbers 6/1 and 7/1 and use addition in $\mathbb{Q}$. The answers are identical since we identify the integer 13 with the rational number 13/1.

Multiplication behaves similarly:

$$(6)(7) = 42$$

and

$$\left(\frac{6}{1}\right)\left(\frac{7}{1}\right) = \frac{42}{1}$$

but

$$\frac{42}{1} = 42$$

so the answers are identical.

Technically, the rational number $n/1$ represents a pair of integers $(n, 1)$ conveniently written in the familiar fraction format. This means that the rational number $n/1$ and the integer n are not identical mathematical objects. However, the identification $n = n/1$ and the resulting consistency of operations allow us to consider them as being the same. In the future, we will make this identification back and forth whenever it seems convenient. Thus we think of the integers as a proper subset of the rational numbers. ($\mathbb{Z}$ is a proper subset since not every rational number is an integer, e.g., 13/15.) Hence,

$$\mathbb{Z} \subset \mathbb{Q} \text{ but } \mathbb{Z} \neq \mathbb{Q}.$$

In particular 0 and 1 may be considered as elements of $\mathbb{Q}$: 0 = 0/1, 1 = 1/1. Since these integers act (by theorem 7) as additive and multiplicative identities, respectively, for $\mathbb{Z}$, we might expect them also to play these roles for the larger set $\mathbb{Q}$. They do. The proof is easy:

Let $\dfrac{n}{m} \in \mathbb{Q}$. Then

$$\frac{n}{m} + 0 = \frac{n}{m} + \frac{0}{1}$$

$$= \frac{n \cdot 1 + n \cdot 0}{m \cdot 1}$$

$$= \frac{n}{m}.$$

Also,

$$\frac{n}{m} \cdot 1 = \frac{n}{m} \cdot \frac{1}{1}$$

$$= \frac{n \cdot 1}{m \cdot 1}$$

$$= \frac{n}{m}.$$

Since addition and multiplication are both commutative on $\mathbb{Q}$ (this follows easily from the definitions) we can summarize these two little calculations by:

Let $r \in \mathbb{Q}$. Then

$$r + 0 = 0 + r = r$$

and

$$r \cdot 1 = 1 \cdot r = r.$$

(Remember: if $r \in \mathbb{Q}$ then $r = \dfrac{n}{m}$ for some $m, n \in \mathbb{Z}$ and $m \neq 0$.)

Let's notice in passing that 0 and 1 may also be written as

$$0 = \frac{0}{1} = \frac{0}{n}$$

and

$$1 = \frac{1}{1} = \frac{n}{n}$$

for any $n \in \mathbb{Z}$ with $n \neq 0$. ($\frac{0}{1} = \frac{0}{n}$ because $0 \cdot n = 1 \cdot 0$ and $\frac{1}{1} = \frac{n}{n}$ because $1 \cdot n = n \cdot 1$.)

The order notions $<$ and $\leq$ may also be extended into $\mathbb{Q}$ in such a manner as to be consistent with the usual ordering on the integers. Once this has been done (we omit the details) it follows that, *for particular pairs of rational numbers having equal denominators*, we have

(I)
$$\frac{a}{b} < \frac{c}{b} \Leftrightarrow a < c$$

where the right-hand inequality involves the ordinary "less than" notion for integers. In particular,

$$\frac{6}{7} < \frac{9}{7}$$

because $6 < 9$. Similarly,

(J)
$$\frac{a}{b} \leq \frac{c}{b} \Leftrightarrow a \leq c.$$

These order results (I) and (J) seem restrictive since they apply only to rational numbers having identical denominators. But we can extend them by a "common denominator" technique so that they apply to arbitrary pairs of rational numbers. As an example consider the rational numbers $2/3$ and $12/15$. Which is larger? (Notice they are not equal since $(2)(15) \neq (3)(12)$, i.e., $30 \neq 36$.)

Write

$$\frac{2}{3} = \frac{2}{3} \cdot 1$$

$$= \frac{2}{3} \cdot \frac{5}{5} \quad (1 = \frac{n}{n} \text{ for any integer } n \neq 0)$$

$$= \frac{2 \cdot 5}{3 \cdot 5}$$

$$= \frac{10}{15}.$$

But (I) now says

$$\frac{10}{15} < \frac{12}{15}$$

because

$$10 < 12.$$

Hence,

$$\frac{2}{3} < \frac{12}{15}.$$

As a second example consider $\frac{-6}{13}$ and $\frac{-5}{12}$. Then,

$$\frac{-6}{13} = \frac{-6}{13} \cdot 1$$

$$= \frac{-6}{13} \cdot \frac{12}{12}$$

$$= \frac{-72}{156}$$

and

$$\frac{-5}{12} = \frac{-5}{12} \cdot 1$$

$$= \frac{-5}{12} \cdot \frac{13}{13}$$

$$= \frac{-65}{156}.$$

Thus,

$$\frac{-6}{13} < \frac{-5}{12}$$

because

$$-72 < -65.$$

The trick in this comparison calculation stems from the observation that, in general, the rational numbers n/m and p/q can be replaced by equivalent rational numbers having the same denominator by multiplying the first by $1 = q/q$ and the second by $1 = m/m$. This gives

$$\frac{n}{m} = \frac{n}{m} \cdot 1$$

$$= \frac{n}{m} \cdot \frac{q}{q}$$

$$= \frac{nq}{mq}$$

and

$$\frac{p}{q} = \frac{p}{q} \cdot 1$$

$$= \frac{p}{q} \cdot \frac{m}{m}$$

$$= \frac{p \cdot m}{qm}.$$

Then (I) tells us that $\dfrac{nq}{mq} < \dfrac{pm}{mq}$ if and only if $nq < pm$.

On $\mathbb{Q}$, exactly as in the case of $\mathbb{Z}$, we write $s < t \Leftrightarrow t > s$. (For example, $\dfrac{9}{7} > \dfrac{6}{7}$ because $\dfrac{6}{7} < \dfrac{9}{2}$.) Thus any of the above inequalities could be reversed provided we replace "<" with ">."

We could now proceed, with only a modicum of effort, to establish the fundamental properties of addition, multiplication, and order on $\mathbb{Q}$. Once the basic definitions have been formulated the course of development of the rational numbers, unlike true love, runs smooth. However, except for what we have already done and for that left to the exercises, we will omit the details. Instead, we will simply accept the following theorem that describes the basic results for $\mathbb{Q}$ as we did its companion (theorem 7) in regard to the integers:

THEOREM 16. Let $\mathbb{Q}$ denote the set of rational numbers. Then:

A   (i) $x, y \in \mathbb{Q} \Rightarrow (x + y) \in \mathbb{Q}$ ($\mathbb{Q}$ is closed with respect to addition)

    (ii) $x, y \in \mathbb{Q} \Rightarrow x + y = y + x$ (Addition of integers is *commutative*)

    (iii) $x, y, z \in \mathbb{Q} \Rightarrow x + (y + z)$ (Addition is *associative*)
         $= (x + y) + z$

    (iv) $x \in \mathbb{Q} \Rightarrow x + 0 = 0 + x = x$ (0 is an additive identity for $\mathbb{Q}$)

    (v)   If $x \in \mathbb{Q}$ then there exists   (w is an additive inverse
       $w, \in \mathbb{Q}$ such that   for x. We write $w = -x$)
       $x + w = w + x = 0$

M   (i) $x, y \in \mathbb{Q} \Rightarrow x \cdot y \in \mathbb{Q}$ ($\mathbb{Q}$ is *closed* with respect to multiplication)

    (ii) $x, y \in \mathbb{Q} \Rightarrow x \cdot y = y \cdot x$ (Multiplication is commutative)

    (iii) $x, y, z \in \mathbb{Q} \Rightarrow x \cdot (y \cdot z)$ (Multiplication is associative)
         $= (x \cdot y) \cdot z$

    (iv) $x \in \mathbb{Q} \Rightarrow x \cdot 1 = 1 \cdot x = x$ (1 is a multiplicative identity for $\mathbb{Q}$)

    (v)   If $x \in \mathbb{Q}$ and $x \neq 0$ then   (v is a multiplicative inverse
       there exists $v, \in \mathbb{Q}$ with   for x. We write $v = x^{-1}$)
       $x \cdot v = v \cdot x = 1$

D   (i) $x, y, z \in \mathbb{Q} \Rightarrow x(y + z)$ (Multiplication is distributive
       $= xy + xz$ with respect to addition)

O   (i)   If $x, y, z \in \mathbb{Q}$ then $x > y$ ($>$ is transitive)
       and $y > z \Rightarrow x > z$

    (ii)   If $x, y, z \in \mathbb{Q}$ then   ($>$ is preserved by addition)
       $x > y \Rightarrow x + z > y + z$

    (iii)   If $x, y, z \in \mathbb{Q}$ and $z > 0$ ($>$ is preserved by multiplication
       then $x > y \Rightarrow xz > yz$ by positive rational numbers)

    (iv)   If $x, y, z \in \mathbb{Q}$ and $z > 0$ (A *positive rational number* may be
       then $xz > yz \Rightarrow x > y$ cancelled from each side of an
                                   inequality)

The striking similarity with theorem 7 might have been expected. Theorem 7 describes the basic properties of addition and multiplication for the set of integers and theorem 16 does the same for the rational numbers. But the integers (via the identification $n = n/1$) form a subset of the rational numbers. There would be little point in defining operations of addition and multiplication on the larger set unless they were chosen so that they inherit the basic properties they possess on the smaller set. It is precisely the desire that these properties carry over to $\mathbb{Q}$ that led earlier mathematicians to define addition and multiplication in the manner we have seen:

(C)
$$\frac{m}{n} + \frac{p}{q} = \frac{mq + np}{nq} \text{ and } \frac{m}{n} \cdot \frac{p}{q} = \frac{mp}{nq}.$$

At the outset, for example, we *might* have defined addition on $\mathbb{Q}$ as

(W)
$$\frac{m}{n} + \frac{p}{q} = \frac{m + p}{n + q}$$

at least for the case $n + q \neq 0$. But this definition does not yield the proper additive properties on the integers. In particular it says:

$$1 + 2 = \frac{1}{1} + \frac{2}{1} = \frac{3}{2}.$$

Definition (W) also gives

$$\frac{1}{2} + \frac{1}{2} = \frac{1+1}{2+2} = \frac{2}{4} = \frac{2 \cdot 1}{2 \cdot 2} = \frac{1}{2}.$$

These absurd results tell us that definition (W) would have been useless. Consequently, *we did not use it as the meaning of addition of rational numbers.* Take one good look at (W), understand why it fails, and then *banish it forever from your memory.* The correct definitions are given in (C). Definition (W) is wrong.

Although the elements of $\mathbb{Q}$ are indicated in theorem 15 by single symbols such as $x, y, \ldots, z$, you must remember that each rational number is actually a pair of integers written in the form of a fraction like $x = n/m$ (with $m \neq 0$). This notation makes the similarity with theorem 7 vivid. But, in the interpretation of the results in theorem 15, we will often go back to the "fractional notation." We particularly need fractional notation when we look at part M(v), which has no counterpart in theorem 7.

As in theorem 7, the divisions of the results of theorem 16 are broken into categories labeled A, M, D, and O. Once again, A stands for "addition," M for "multiplication," D for "distributive," and O for "order." The parenthetic remarks provide convenient phrases we may associate with each result in a manner identical with theorem 7. Part M(v) of the present theorem speaks of the notion of "multiplicative inverse" for each nonzero rational number.

You will recall that no such integer exists for a given integer n. It was, in fact, mainly the absence of multiplicative inverses that led to the construction of the rational numbers. Part M(v) ensures that—whatever the rational numbers may lack in terms of desirable properties—they do not have this particular deficiency.

According to M(iv), the rational number 1 serves as a multiplicative identity in $\mathbb{Q}$. But the rational number 1 is also the integer $1(1 = 1 / 1)$, which by theorem 7 is a multiplicative identity for $\mathbb{Z}$. When we discussed theorem 7, it was pointed out that the integer 1 *stands as the only multiplicative identity for* $\mathbb{Z}$. Similarly, 1 *is the unique multiplicative identity for* $\mathbb{Q}$. Here is the proof:

Suppose there is another multiplicative identity, say u, for $\mathbb{Q}$. Then

$$1 \cdot u = u \text{ (because 1 is the multiplicative identity).}$$

Similarly,

$$1 \cdot u = 1 \text{ (because } u \text{ is a multiplicative identity).}$$

Thus,

$$1 = u.$$

So the multiplicative identity is unique.

Thus, we may replace the word "a" by "the" in the parenthetic remark in M(iv) and speak of "the multiplicative identity in $\mathbb{Q}$." Similarly, we may refer to "the additive inverse for $\mathbb{Q}$," since there is only one, namely, the rational number 0.

Also, the multiplicative inverse, guaranteed for each $x \neq 0$ by M(v), is unique. The proof goes like this:

Let $x \in \mathbb{Q}$, $x \neq 0$. Suppose x has two multiplicative inverses, u and v. Then

$$x \cdot u = 1 \text{ and } x \cdot v = 1.$$

Hence

$$x \cdot u = x \cdot v$$

so that

$$u \cdot (x \cdot u) = u \cdot (x \cdot v).$$

Then

$$(u \cdot x) \cdot u = (u \cdot x) \cdot v$$
$$1 \cdot u = 1 \cdot v.$$

So,

$$u = v.$$

Notice that the word "unique" has one interpretation in M(iv) and another in M(v). To say that $\mathbb{Q}$ has a unique multiplicative identity means that there exists a single rational number, namely, 1, such that $1 \cdot x = x \cdot 1 = x$ for each $x \in \mathbb{Q}$. To say that multiplicative inverses are unique means that, for each $x \in \mathbb{Q}$, there exists a single rational number v such that $x \cdot v = v \cdot x = 1$. The unique number 1 works for all x's. But as we shall see, the inverse v depends on the rational number x.

In fact, the multiplicative inverse of the rational number $x = \dfrac{n}{m}$ is the number $v = m/n$. Since we know that multiplicative inverses are unique, the proof is easy.

Let $x = n/m$ and $v = m/n$. (Of course, $n, m \in \mathbb{Z}$ and $n \neq 0, m \neq 0$.) Then

$$x \cdot v = \left(\frac{n}{m} \right) \left(\frac{n}{m} \right)$$
$$= \frac{nm}{mn}$$
$$= \frac{mn}{mn}$$
$$= 1.$$

Thus v is *the* multiplicative inverse for x. (We need not bother to show $v \cdot x = 1$ since A(ii) ensures us that $v \cdot x = x \cdot v$.)

The parenthetic remark accompanying M(v) says that we write $v = x^{-1}$.

Hence

$$\left(\frac{n}{m}\right)^{-1} = \frac{m}{n}.$$

So, *there is no trick to finding the multiplicative inverse of a fraction that is not zero: just turn it upside down.*

We now must give some attention to this matter of zero with respect to rational numbers. Two questions occur:

(i) What is the reason for the restriction $m \neq 0$ in the definition of a rational number x as $x = n/m$?

(ii) Why does the existence of a multiplicative inverse for a rational number x in M(v) require that $x \neq 0$?

These questions can be most easily settled with the use of the two following theorems:

THEOREM 17. Let a and b be integers with $b \neq 0$. Then

$$\frac{a}{b} = r \Leftrightarrow a = br.$$

PROOF. First suppose $\frac{a}{b} = r$ .Then $b \cdot \frac{a}{b} = br$. So,

$$\frac{ba}{b} = br$$

and cancellation gives

$$a = br.$$

Thus,

$$\frac{a}{b} = r \Rightarrow a = br.$$

Now suppose $a = br$. Since $b \neq 0$, b^{-1} exists. Then

$$b^{-1}a = b^{-1}(br)$$
$$= (b^{-1}b)r$$
$$= 1 \cdot r$$
$$= r.$$

But $b^{-1} = \dfrac{1}{b}$. So $\left(\dfrac{1}{b}\right) \cdot a = r$, or $\dfrac{a}{b} = r$. Then, $a = br \Rightarrow \dfrac{a}{b} = r$. Consequently,

$$\frac{a}{b} = r \Leftrightarrow a = br.$$

THEOREM 18. Let r be any rational number. Then

$$0 \cdot r = 0.$$

PROOF. The proof is formally identical to that of the corresponding theorem for integers (theorem 12(A)) and is left as an exercise.

Now go back to question (i). Let $x = n/m$ be any rational number. Theorem 16 says

$$x = \frac{n}{m} \Leftrightarrow mx = n.$$

So, if $m = 0$ then

$$x = \frac{n}{0} \Leftrightarrow 0 \cdot x = n.$$

But $0 \cdot x = 0$. Hence,

$$x = \frac{n}{0} \Leftrightarrow n = 0.$$

Consequently, if we were to allow $m = 0$ in the rational number expression $x = n/m$, it follows that we must also have $n = 0$. This gives $x = 0/0$. But theorem 17 then says

$$x = \frac{0}{0} \Leftrightarrow x \cdot 0 = 0.$$

But $x \cdot 0 = 0$ is true for each rational number x by theorem 17. Hence, x does not represent a well-defined rational number.

This discussion shows that we cannot have $m = 0$ in any expression $x = n/m$. Sometimes mathematicians distinguish between the case $x = n/0$ when $n \neq 0$ and the case $x = 0/0$. The first expression they call *undefined* and the second *indeterminate. Both, however, are improper and although we had to write n/0 and 0/0 for purposes of illustration, it is best not to write such expressions at all.*

The answer to question (ii) follows easily from theorem 18. Since $r \cdot 0 = 0$ for any rational number r we can never have $v \cdot 0 = 1$ for $v \in \mathbb{Q}$. Thus 0 has no multiplicative inverse.

MANIPULATIONS

If you think of the rational number n/m as a fraction, then you may want to use the school language "a/b is the integer a divided by the integer b." There is nothing wrong with this except that it is unnecessary since the rational numbers have been defined without any reference to the notion of division. However, we can enlarge our set of symbols of the form x/y (not our set of numbers) and simultaneously introduce a division concept.

DEFINITION 3. Let x and y be rational numbers with $y \neq 0$. We write

$$z = \frac{x}{y}$$

and say that z represents x divided by y if and only if

$$yz = x.$$

Theorem 17 tells us that this property holds whenever x, y, and z are integers with $y \neq 0$ so that the new definition remains consistent with what has gone before. However, definition 3 allows us now to write more complicated expressions of the form xy and give to them precise meaning.

For example, we may write

$$2 = \frac{\dfrac{1}{2}}{\dfrac{1}{4}}$$

since $\dfrac{1}{4} \cdot 2 = \dfrac{1}{2}$. (Here, $x = \dfrac{1}{2}$, $y = \dfrac{1}{4}$, $z = 2$.)

You see immediately that the restriction $y \neq 0$ in definition 3 is necessary. The necessity follows in formal manner exactly as the discussion following theorem 17. No meaning can be assigned to the symbol x/y when $y = 0$, even when x may be an arbitrary rational number. (The earlier discussion gave the same conclusion in the case where x was an arbitrary integer.) All together this results in the iron rule:

Division by zero is not allowed.

Never, never, attempt to divide by zero. Trouble always follows—often a sea of troubles. (We will see an example at the end of the chapter.)

We are now in position to formulate a number of rules that allow the quick and proper manipulation of symbols of the form x/y, where x and y are rational numbers. (These rules compose the bulk of the subject taught in secondary schools under the name "algebra.") Most of the rules appear in theorem 16 and the other results of this chapter. But additional rules follow from these. For example, definition 3 tells us that

$$\frac{\dfrac{a}{b}}{\dfrac{c}{d}} = r \Leftrightarrow \frac{a}{b} = \frac{c}{d} \cdot r.$$

But this gives

$$\frac{a}{b} \cdot d \cdot c^{-1} = \frac{c}{d} \cdot r \cdot d \cdot c^{-1}$$

or

$$\frac{a}{b} \cdot d \cdot \frac{1}{c} = \frac{c \cdot c^{-1}}{d} \cdot d \cdot r$$

or

$$\frac{a}{b} \cdot \frac{d}{c} = r.$$

(In particular, we have used $c^{-1} = \dfrac{1}{c}$, $c \cdot c^{-1} = 1$, $\dfrac{d}{d} = 1$.) Setting equal the two expressions for r gives,

$$\frac{\dfrac{a}{b}}{\dfrac{c}{d}} = \frac{a}{b} \cdot \frac{d}{c}$$

and we have the familiar rule:

In order to divide one fraction by another, invert the fraction in the denominator and multiply.

In particular:

$$\frac{\dfrac{1}{2}}{\dfrac{1}{4}} = \frac{1}{2} \cdot \frac{4}{1} = \frac{4}{2} = 2.$$

We can also exploit the "common denominator" technique used earlier for comparing sizes of rational numbers to obtain a useful method for adding fractions. It goes like this:

$$\frac{a}{b} + \frac{c}{d} = \frac{a}{b} \cdot 1 + \frac{c}{d} \cdot 1$$

$$= \frac{a}{b} \cdot \frac{d}{d} + \frac{c}{d} \cdot \frac{b}{b}$$

$$= \frac{ad}{bd} + \frac{cd}{bd}$$

$$= \frac{ad + cd}{bd}.$$

The trick here, of course, is to multiply each fraction in the sum by 1. But the fractional representation of 1 is chosen, in each case, so that common denominators result. Nothing is gained when the technique is applied to the sum of two ordinary rational numbers since the definition of addition yields the identical answer more quickly. But the technique applies also to fractional sums, $\dfrac{a}{b} + \dfrac{c}{d}$, when a, b, c, d may themselves be complicated combinations of rational numbers. (See later example E.) Moreover, the technique easily extends to sums having more than two terms. For example,

$$\frac{1}{2}+\frac{1}{3}+\frac{1}{4}=\frac{1}{2}\cdot\frac{3}{3}\cdot\frac{4}{4}+\frac{1}{3}\cdot\frac{2}{2}\cdot\frac{4}{4}+\frac{1}{4}\cdot\frac{2}{2}\cdot\frac{3}{3}$$

$$=\frac{12}{24}+\frac{8}{24}+\frac{6}{24}$$

$$=\frac{26}{24}$$

$$=\frac{13}{12}.$$

(Note: $\frac{13}{12}=1+\frac{1}{12}$, which is sometimes written as $1\frac{1}{12}$. But the form $\frac{13}{12}$ is preferable and representations such as $1\frac{1}{12}$ almost never appear in mathematics.)

The process of symbol manipulation goes easier if we introduce a notion of substitution of rational numbers similar to the concept of integer subtraction. Thus, we write, for $r, s \in \mathbb{Q}$

$$r - s = r + (- s).$$

Expected facts like:

$$r - s = t \Leftrightarrow r = s + t$$

and

$$r > s \Leftrightarrow r - s > 0$$

then follow. (We omit the proof.)

Moreover, theorem 16 then yields for $\mathbb{Q}$ certain consequences analogous to those obtained for $\mathbb{Z}$ from theorem 7. These include:

$$-\left(-\frac{a}{b}\right) = \frac{a}{b},$$

$$\left(-\frac{a}{b}\right)\left(-\frac{c}{d}\right) = \left(\frac{a}{b}\right)\left(\frac{c}{d}\right),$$

$$\frac{-\dfrac{a}{b}}{-\dfrac{c}{d}} = \frac{\dfrac{a}{b}}{\dfrac{c}{d}}$$

$$= \frac{ad}{bc},$$

$$-\frac{a}{b} = \frac{-a}{b} = (-1)\frac{a}{b}, \text{ and}$$

$$x - \left(-\frac{a}{b}\right) = x + \frac{a}{b}.$$

These facts follow straightforwardly and we shall not bother with all the details. As an illustration, here is the proof of the first item:

THEOREM 19. Let $x = a/b$ be a rational number. Then

$$-(-x) = x.$$

PROOF. According to theorem 16, A(v) $w = -x$ is the (unique) rational number such that $x + w = 0$. But this equation also says that x is the additive inverse for w. Thus, $x = -w$. So, $x = -w = -(-x)$.

The list of rules goes on and it is pointless to try for completeness. In any special case requiring simplification or equation solving, the necessary manipulative steps can be *derived* from the definitions and the basic theorem. New rules may always be created from existing rules. But we can do fairly well with what we have already learned. Several examples follow.

EXAMPLE A. Evaluate

$$\frac{\dfrac{1}{2} + \dfrac{1}{3}}{\dfrac{1}{12}}.$$

SOLUTION.

$$\frac{\dfrac{1}{2}+\dfrac{1}{3}}{\dfrac{1}{12}}=\frac{\dfrac{3}{6}+\dfrac{2}{6}}{\dfrac{1}{12}}=\frac{\dfrac{5}{6}}{\dfrac{1}{12}}$$

$$=\frac{5}{6}\cdot\frac{12}{1}=\frac{5}{6}\cdot 6\cdot 2$$

$$=5\cdot 2=10.$$

EXAMPLE B. Evaluate

$$\frac{\dfrac{-4}{-2-(-6)}}{\dfrac{2(-2)}{3(-2+6)}}.$$

SOLUTION.

$$\frac{\dfrac{-4}{-2-(-6)}}{\dfrac{2(-2)}{3(-2+6)}}=\frac{\dfrac{-4}{-2+6}}{\dfrac{-4}{3\cdot 4}}=\frac{\dfrac{-4}{4}}{\dfrac{1}{3}}$$

$$=\frac{-1}{-\dfrac{1}{3}}=(-1)\left(-\frac{3}{1}\right)=3.$$

EXAMPLE C. Let $n, m \in \mathbb{Z}$. Write

$$\frac{\dfrac{n}{n+1}-\dfrac{m}{m+1}}{1-\dfrac{m+1}{n+1}}$$

in the form p/q.

SOLUTION.

$$\frac{\dfrac{n}{n+1}-\dfrac{m}{m+1}}{1-\dfrac{m+1}{n+1}}=\frac{\left(\dfrac{n}{n+1}\right)\left(\dfrac{m+1}{m+1}\right)-\left(\dfrac{m}{m+1}\right)\left(\dfrac{n+1}{n+1}\right)}{\left(\dfrac{n+1}{n+1}\right)-\left(\dfrac{m+1}{n+1}\right)}$$

$$=\frac{\dfrac{n(m+1)-m(n+1)}{(n+1)(m+1)}}{\dfrac{(n+1)-(m+1)}{(n+1)}}$$

$$=\frac{(nm+n-mn-m)}{(n+1)(m+1)}\cdot\frac{(n+1)}{(n+1-m-1)}$$

$$=\frac{(n-m)}{(n+1)(m+1)}\cdot\frac{(n+1)}{(n-m)}$$

$$=\frac{1}{m+1}.$$

(Notice that we must have $n \neq -1$, $m \neq -1$, $n \neq m$ in order to ensure nonzero denominators.)

EXAMPLE D. Solve the equation $-3x + 2 = 0$.

SOLUTION.

$$-3x + 2 = 0 \Rightarrow$$
$$-3x = -2 \Rightarrow$$
$$(-3)x = -2 \Rightarrow$$
$$x = \frac{(-2)}{(-3)} = \frac{2}{3}.$$

(This is a special case of the general linear equation $ax + b = 0$. If $a \neq 0$, the solution is $x = -b/a$.)

EXAMPLE E. Solve the equation $\dfrac{1}{5x} - x = \dfrac{2 - 15x}{15}$.

SOLUTION.

$$\frac{1}{5x} - x = \frac{2 - 15x}{15} \Rightarrow$$

$$x\left(\frac{1}{5x} - x\right) = x\left(\frac{2 - 15x}{15}\right) \Rightarrow$$

$$\frac{x}{5x} - x^2 = \frac{2x - 15x^2}{15} \Rightarrow$$

$$15\left(\frac{1}{5} - x^2\right) = 2x - 15x^2 \Rightarrow$$

$$3 - 15x^2 = 2x - 15x^2 \Rightarrow$$

$$3 = 2x \Rightarrow$$

$$x = \frac{3}{2}.$$

(Here we have used the standard notation $x \cdot x = x^2$.)

Before we close this chapter, let's look more carefully at the notation $x \cdot x = x^2$ of the last example. This generalizes naturally to

$$x = x^1$$

$$x^2 = x \cdot x$$

$$x^3 = x \cdot x \cdot x$$

$$\cdots$$

$$x^n = x \cdot x \cdot x \cdot \ldots x.$$

when the last expression has n terms on the right-hand side. It is convenient to define $x^0 = 1$ and negative powers of x by

$$x^{-n} = \frac{1}{x^n}$$

for $n = 1, 2, 3, \ldots$.

From this notation and the properties of rational numbers we obtain for $x, y \in \mathbb{Q}$:

$$x^n \cdot x^m = x^{n+m},$$

$$x^n / x^m = x^{n-m}, \text{ and}$$

$$\left(x^n\right)^m = x^{mn},$$

which are the familiar "laws of exponents." (The laws of exponents follow easily from the definitions exactly as does this special case:

$$\frac{x^5}{x^3} = \frac{x \cdot x \cdot x \cdot x \cdot x}{x \cdot x \cdot x} = x \cdot x = x^2 = x^{5-3}.$$

We will use exponent notation extensively when we come to more involved mathematical analysis. At the moment, I only want to point out two useful identities:

$$(x + y)^2 = x^2 + 2xy + y^2$$

and

$$(x - y)(x + y) = x^2 - y^2.$$

The proof of the second identity is:

$$(x - y)(x + y) = (x + (-y))(x + y)$$
$$= x(x + y) + (-y)(x + y)$$
$$= x^2 + xy - xy - y^2$$
$$= x^2 - y^2.$$

We can use the second identity to illustrate the "sea of troubles" that arises whenever you divide by zero. Kronecker believed that God gave us the natural numbers. Maybe. I don't know. But I am ready to believe that he did give us an eleventh commandment:

Thou shalt not divide by zero.

Violate this and you die.
Now watch me while I sin:

Let x and y be nonzero rational numbers. Set $x = y$. Then

$$x = y \Rightarrow x^2 = xy$$
$$\Rightarrow x^2 - y^2 = xy - y^2$$
$$\Rightarrow (x - y)(x + y) = y(x - y)$$
$$\Rightarrow x + y = y$$
$$\Rightarrow x + x = x$$
$$\Rightarrow 2x = x$$
$$\Rightarrow 2 = 1.$$

Thus $2 = 1$. If this be true and upon me proved then I have plunged us into the troubled sea. For then $1 + 1 = 2 + 1$, or $2 = 3$. And then $3 = 4$, $4 = 5$, and on and on until all the numbers God gave Kronecker become identical to one another. But also $2 = 1$ gives $2 - 1 = 1 - 1$ or $1 = 0$. Thus all the natural numbers become equal, and equal to zero.

The cold sea rises. And all becomes zero, naught, nothing—all this, if 1 be nothing.

> Why, then the world and all that's in it is nothing;
> The covering sky is nothing; Bohemia nothing;
> My wife is nothing; not nothing have these nothings,
> If this be nothing.[3]

Chapter 4

NUMBER THEORY

An often-told tale bears repeating.

Carl Friedrich Gauss was ten years of age in 1787. At about that time he entered a class in arithmetic. One day the teacher assigned the students this problem: find the sum of all the natural numbers from 1 to 100. Each student—except Gauss—laboriously began to compute the sum $1 + 2 + 3 + 4 + \ldots + 100$. Gauss sat still and thought. He noticed that the numbers $1, 2, 3, \ldots, 100$ could be neatly written in the double column

1	100
2	99
3	98
⋮	⋮
49	52
50	51

The rows then yielded the pairings $(1, 100), (2, 99), \ldots, (50, 51)$. The numbers in each pair added to 101 and there are exactly 50 pairs. Thus the sum $1 + 2 + \ldots + 100$ equals $(50) \cdot (101) = 5{,}050$. All this, Gauss did in his head. In seconds. He wrote nothing on his slate except the answer: 5,050. "There it lies," the kid said to his teacher.

The teacher—who had no inkling of his pupil's thought process—believed that Gauss had guessed.

This schoolboy, Gauss, grew up to become the person considered by many to be the greatest mathematician of them all. It is no cliché to say that Gauss's contributions to mathematics are too numerous to mention. No specialist he, this man mastered all of the mathematics of his time and extended most of it. He produced seminal work in number theory, real analysis, complex analysis, special functions, and differential geometry. His interests—

and his research—included both the practical and the mathematical aspects of astronomy and physics. George Simmons calls Gauss the "supreme mathematician" and says of him:

> He surpassed the levels of achievement possible for ordinary men of genius in so many ways that one sometimes has the eerie feeling that he belonged to a higher species.[1]

Without Gauss—or someone like him—the phrase "ordinary men of genius" would be an oxymoron.

Gauss's first masterpiece (perhaps his greatest) was the famous *Disquisitones Arithmeticae*. In this book of seven sections—the book of seven seals—Gauss formalized and made rigorous Fermat's subject of number theory. The *Disquisitones* gave number theory the same unified status as the other existing areas of mathematics. The book turned the isolated fragments left by Fermat and others into a true subject. (Newton and Leibniz had earlier done the same for calculus.)

In the *Disquisitones*, Gauss brought to bear on number theory the powerful analytic techniques available through the use of the theory of complex variables. This produced many results that would otherwise be unreachable and showed, for the first time, the deep and fundamental connections between the smooth, continuous world of analysis and the discrete, bumpy domain of the integers. We cannot go nearly so far here. Gauss referred to number theory as "the higher arithmetic" and we will confine ourselves to only a few basic results of arithmetic type. Afterward we will examine briefly some natural abstractions of these arithmetic properties that lead into another branch of mathematics: the branch called *algebra*.

PRIMES

Consider the natural number 30. Elementary arithmetic tells us that $30 = 6 \cdot 5$. But $6 = 2 \cdot 3$. Consequently, we may write

$$30 = 2 \cdot 3 \cdot 5.$$

In the schools this process is called *factoring*. We began with the number 30 and then "factored" it into "six times five." Next, we factored 6 into "two times three." This enabled us to write $30 = 2 \cdot 3 \cdot 5$. But the process ends here. None of the numbers 2, 3, or 5 may be factored further. Such natural numbers are called *primes*: 2, 3, and 5 are examples of *prime numbers*. The representation $30 = 2 \cdot 3 \cdot 5$ yields a representation of the number 30 as a *product of primes*.

Evidently, we can duplicate this process for any natural number. Consider, for example, the number 385. A trial and error search for *divisors* of 385 gives: $385 = (35)(11)$. But $35 = 5 \cdot 7$. Thus,

$$385 = 5 \cdot 7 \cdot 11$$

and we have written 385 as a product of primes.

The precise statement that such a representation exists for any natural number (larger than 1) is called the *fundamental theorem of arithmetic*. One of our objectives is to produce this theorem and to discuss the surrounding circle of ideas. In order to do so, we must first deal more carefully with the notions of *divisor* and of *prime number*.

DEFINITION 1. Let n and d be nonzero integers. We say that d divides n and we write $d|n$ if and only if there exists an integer q such that $n = dq$.

REMARK. It follows immediately that the integer q—*if* it exists—cannot be zero. (Because $d \cdot 0 = 0$ but $n \neq 0$.) Also $d \neq 0$ so that

$$d \mid n \Leftrightarrow n = dq$$

$$\frac{n}{d} = q.$$

Then, $d|n$ (i.e., d divides n) if and only if the rational number $\frac{n}{d}$ is an integer q.

The symbols d and q are used here to remind us of the familiar terms *division* and *quotient*. If $d|n$ is false (i.e., there is no integer q such that $n = dq$) we may write $d \nmid n$. (Just as $x \neq y$ means "x is not equal to y," $d \nmid n$ means "d does not divide n.")

Notice that each positive integer always has 1 and n as divisors because

$$n = 1 \cdot n.$$

For example, $2 \mid 36$ because $36 = (2)(18)$ or $\frac{36}{2} = 18$. Also, $5 \mid 55$ because $55 = (5)(11)$ or $\frac{55}{5} = 11$.

However, $5 \mid 19$ *is false* because there exists no integer q such that $19 = 5q$, that is, the rational number $\frac{19}{5}$ is not an integer. Thus, $5 \nmid 19$.

The remark following the above definition allows us to relate the notion of "d divides n" to the concept of the rational number $\frac{n}{d}$ being equal to an integer q. This observation turns out to be useful and it is appropriate to make it at this time. However, the notion of $d \mid n$ involves only the integers and is, therefore, more fundamental than ideas associated with the rational numbers $\mathbb{Q}$. In fact, definition 1 involves only multiplication of integers. In this sense, *division is simply restated multiplication.* Thus we could discuss the notion $d \mid n$ without mentioning rational numbers. But, since we have the rationals at hand, we will not hesitate to make use of them.

Two properties of the notion of "divides" follow immediately from the definition.

THEOREM 1.   (i) $d \mid n$ and $d \mid m \Rightarrow d \mid (n + m)$.

(ii) $d \mid n \Rightarrow d \mid nm$ for any integer $m \neq 0$.

PROOF.   (A) $d \mid n \Rightarrow n = dq_1$ for some $q_1 \in \mathbb{Z}$, $q_1 \neq 0$.

$d \mid m \Rightarrow m = dq_2$ for some $q_2 \in \mathbb{Z}$, $q_2 \neq 0$. Hence,

$$n + m = dq_1 + dq_2$$
$$= d(q_1 + q_2).$$

So

$$d \mid (n + m) \text{ by definition 1.}$$

(B) The proof of part (ii) is left as an exercise.

The proof of (i) could easily be given in terms of rational numbers as follows.

$$d \mid n \Rightarrow \frac{n}{d} \text{ is an integer}$$

and

$$d \mid m \Rightarrow \frac{m}{d} \text{ is an integer}$$

thus

$$\frac{n}{d} + \frac{m}{d} \text{ is an integer}$$

but

$$\frac{n}{d} + \frac{m}{d} = \frac{n+m}{d}.$$

Hence, $d \mid (n + m)$.

Notice that the same proof will yield:

$$d \mid n \text{ and } d \mid m \Rightarrow d \mid (n - m).$$

Thus, we have

$$d \mid n \text{ and } d \mid m \Rightarrow d \mid (n \pm m).$$

In terms of the "divisor notion" the precise definition of *prime number* is:

DEFINITION 2. An integer p is called prime if and only if

(A) $p \geq 2$ and

(B) $d \mid p, d > 0 \Rightarrow d = 1$ or $d = p$.

An integer larger than 1 *that is not prime is called* composite.

For example, 5 is a prime because it has no positive divisors except itself and 1. (According to definition 1, 5 also has (-1) and (-5) as divisors because $5 = (-1)(-5)$. But negative divisors play no role in the definition of prime numbers.)

Notice that 2 is the only even prime since each even number has 2 as a divisor. The number 1 is excluded from the collection of prime numbers for technical reasons. The *first ten prime numbers* then are:

$$2, 3, 5, 7, 11, 13, 17, 19, 23, 29.$$

(The italics in the preceding sentence anticipate a forthcoming result: *there exist infinitely many primes.* So, if we want the above list to represent all the primes, and not simply the first ten, we must place a comma and then three dots after the number 29.)

SIEVE OF ERATOSTHENES

Even though—as we shall see—there exist infinitely many primes, it may be enormously difficult to determine whether or not any particular natural number is prime. If, say, we are given a natural number $n \geq 2$, we may determine its prime status by the obvious—and perhaps painful—method of testing each possible divisor of n. If we proceed naively and inefficiently, this requires testing one by one the numbers $2, 3, \ldots, n - 1$ as possible divisors of n. If we are lucky, we may find a divisor early on. In this case, the process ends and we conclude that n is not prime. However, if n is prime— and if we fail to take advantage of procedural shortcuts—we will check each number $2, 3, \ldots, n - 1$ as a possible divisor of n. This naive method requires $n - 2$ steps. (The list $2, 3, \ldots, m$ contains $m - 1$ numbers.)

For example, consider $n = 79$. Here, the possible divisors are $2, 3, 4, \ldots,$ 78. If we check them one by one and if each step takes 1 minute to complete, the entire procedure will require 77 minutes.

Since $n = 79$ is a "small" number, 77 minutes represents a gross overestimation of the required time. In this case a cursory application of the procedure shows that 79 is a prime number.

On the other hand, matters are not so simple if n is "large." Consider, for example, the natural number $m = 2^{64}$.

Clearly, m *is* not prime because $m = 2 \cdot 2^{63}$ and consequently, m is an even number (larger than 2). But how about the odd number $n = m + 1$? Is n prime?

In this case the naive method requires the testing of each of the possible divisors of

$$n = 2^{64} + 1.$$

Then possible proper divisors are:

$$2, 3, 4, \ldots, 2^{64}.$$

One by one, we test each of these as a possible divisor of n. Perhaps we will find a "small" divisor and end the process quickly. But, if n is prime, and if we are not sufficiently observant to shortcut the work, we will simply plod ahead and check each of the numbers between 2 and 2^{64}. (One obvious shortcut would be to notice that $2^{64} \nmid 2^{64} + 1$.) How much time do we need?

Consider a super-efficient human calculator C who can do this for us at an average check time of one minute per possible divisor. Since there are $2^{64} - 1$ members in the list $2, 3, 4, 5, \ldots 2^{64}$, calculator C will need $t = 2^{64} - 1$ minutes—in the worst case—to do the job. How many years will this require? Let's make a rough estimate of a lower bound. In minutes, the time required is $t = 2^{64} - 1$. But

$$t = 2^{64} - 1$$
$$> 2^{63} \, (Why?)$$
$$= 2^{60+3}$$
$$= 2^{60} \cdot 2^3$$
$$= 8 \cdot 2^{60}$$
$$= 8 \cdot 2^{(4)(15)}$$
$$= 8 \cdot \left(2^4\right)^{15}$$
$$= 8 \cdot (16)^{15}$$
$$> 8 \cdot (10)^{15}.$$

Calculator C, then, needs more than $8 \cdot (10)^{15}$ minutes to do his work. We can convert this estimate to years as follows:

$$t > 8 \cdot (10)^{15} \text{ minutes}$$

$$= 8 \cdot (10)^{15} \text{ minutes} \left[\frac{1 \text{ hour}}{60 \text{ minutes}} \right] \left[\frac{1 \text{ day}}{24 \text{ hours}} \right] \left[\frac{1 \text{ year}}{365 \text{ days}} \right]$$

$$= \frac{8 \cdot (10)^{15}}{(60)(24)(365)} \text{ years.}$$

(This conversion technique works because each of the square-bracketed terms is equal to 1 and the "units"—except for "years"—cancel.) A crude estimate of this gives:

$$t = \frac{8 \cdot (10)^{15}}{(60)(24)(365)} \text{ years}$$

$$= \frac{8 \cdot (10)^{15}}{6 \cdot 10 \cdot 3 \cdot 8 \cdot 365}$$

$$= \frac{(10)^{14}}{6 \cdot 3 \cdot 365}$$

$$> \frac{(10)^{14}}{6 \cdot 3 \cdot 400}$$

$$= \frac{(10)^{14}}{6 \cdot 3 \cdot 4 \cdot (10)^{2}}$$

$$= \frac{(10)^{12}}{72}$$

$$> \frac{(10)^{12}}{100}$$

$$= (10)^{10} \text{ years}$$

$$= 10,000,000,000 \text{ years.}$$

Although this represents a gross underestimation, it still shows that C may need a long time to complete his work. And, even if we replaced C by a computer that could do the calculations a million times faster, the worst-case scenario still requires at least

$$T = \frac{10^{10}}{10^{6}} = 10^{4}$$

$$= 10,000 \text{ years.}$$

Even the best-built computer will not last so long.
Incidentally, my pocket calculator gives the actual values:

$$t = 35,096,540,000,000 \text{ years}$$

and

$$T = 35,096,540 \text{ years.}$$

Thus the actual values greatly exceed the estimated lower bounds. But the rough estimates give helpful information and require no use of modern technology.

In truth, the worst-case scenario does not apply to the number $n = 2^{64} + 1$; C can complete his work in a normal lifetime. Our number n is an example of a *Fermat number*. These numbers are given by

$$F_m = 2^{(2^m)} + 1, \ m = 0, \ 1, \ 2, \ 3, \ \ldots .$$

When $m = 6$, we have

$$F_6 = 2^{2^6} + 1$$

or

$$F_6 = 2^{64} + 1.$$

F_6 is known to be composite (nonprime) with the smallest divisor of 274,177. So, if C plods along naively checking division of $F_6 = n$, he will discover that n is nonprime in only 274,177 minutes—a mere 6 months. The high-speed computer will merely blink and do the job in less than 2 seconds.

Ordinarily, of course, one does not use the naive method to test for primality. Many special techniques have been devised that are particularly effective when coupled with modern computers. Many of these methods require sophisticated mathematics. But some are elementary. And the most ancient of the simple techniques is called the *sieve of Erathosthenes*, after the Greek mathematician who lived 276–194 BCE.

The sieve method depends on the observation that, if n is composite, then it must have a divisor whose square is no larger than n. Suppose

$$n = ab, \ a,b \in \mathbb{N}.$$

Then

$$n^2 = a^2 b^2.$$

Thus, $a^2 \le n$ or $b^2 \le n$ must hold. (Since otherwise $a^2 > n$ and $b^2 > n$ so that $a^2 b^2 > n \cdot n = n^2$.) Consequently, a composite number n, must have a divisor d with $d^2 \le n$.

It would be convenient here—and conventional—to say simply: "n must have a divisor d with d no larger than the square root of n; that is, $d \le \sqrt{n}$, where $\sqrt{n}$ is the usual school symbol for the square root of n. However, we have not yet defined $\sqrt{n}$ and—as we shall soon learn from Pythagoras—$\sqrt{n}$ may not belong to the set of numbers we presently have at our disposal. That is, $\sqrt{n}$ may not be a rational number. However, no harm will come if we anticipate the construction of the real numbers and think of $\sqrt{n}$ as being defined by $\left(\sqrt{n}\right)^2 = n$. Then we can remember Eratosthenes' sieve as the statement:

A composite natural number n *must have a divisor* d *with* $d \le \sqrt{n}$.

For example, look back at $n = 79$. Notice that $8^2 = 64 < 79$ and $9^2 = 81 > 79$. Hence $\sqrt{79}$ falls somewhere between 8 and 9: $8 < \sqrt{79} < 9$. ($\sqrt{79}$ is not a rational number.) The largest integer that does not exceed $\sqrt{79}$ is, therefore, 8. So, in order to determine whether or not 79 is prime, we need test only the possible divisors 2, 3, 4, 5, 6, 7, 8. None of these divide 79 so we see quickly that 79 is prime.

MATHEMATICAL INDUCTION

Our earlier statement of the fifth Peano axiom involves the concept of the *successor* n' of a natural number n. However, we now know that $n' = n + 1$ for each $n \in \mathbb{N}$. This allows us to write the fifth axiom in a slightly altered form. It then becomes:

The principle of mathematical induction: Let $S \subset \mathbb{N}$. If

- $1 \in S$, and
- $n \in S \Rightarrow (n+1)\ S$, then

$$S = \mathbb{N}.$$

In this form the principle of induction remains equivalent to the fifth Peano axiom have simply replaced the notion of successor with the concept of addition. So the analogy with the infinite train of falling dominos remains valid. Another apt analogy involves a long ladder—like the one Jacob saw in his dream.

Consider a ladder with as many rungs as there are natural numbers. You climb it according to the principle of mathematical induction. Let R denote the collection of rungs you climb. Statement (1) of the induction principle ensures that you climb the first rung. Statement (2) says that, whenever you climb any particular rung, you will also climb the next one. The conclusion of the induction principle ensures that $R = \mathbb{N}$. Thus you climb them all. You may climb all the way to heaven if you wish, like the angels near Beersheba. But be careful. Like Robert Frost's swinger of birches,[2] you may wish to return. In that case, you are on your own. Mathematical induction says nothing about descent.

In this new form, the principle of mathematical induction provides a powerful tool for establishing results that hold for each natural number. Consider, for example, the formula

(a) $$1+2+3+4+ \ldots +n = \frac{n(n+1)}{2}.$$

An easy check shows that the formula holds for $n = 1, 2, 3, 4$. (When $n = 4$, we have $1+2+3+4 = 10 = \dfrac{20}{2} = \dfrac{4 \cdot 5}{2}$.) When $n = 100$ the formula gives

$$1+2+3+4+ \ldots +100 = \frac{(100)(101)}{2}$$
$$= 5,050$$

a result we know to be correct because of Gauss's school exercise.

These checks lead to the *conjecture* that the formula holds for each $n \in \mathbb{N}$. Such is the case and we prove it by mathematical induction.

Let $S = \left\{ n : n \in \mathbb{N} \text{ and } 1 + 2 + 3 + \ldots + n = \dfrac{n(n+1)}{2} \right\}$. ($S$ is the set of natural numbers for which the formula holds.)

(1) $1 \in S$, because $1 = \dfrac{1 \cdot (1 = 1)}{2}$.

(2) Suppose $n \in S$. Then

$$1 + 2 + 3 + \ldots + n = \frac{n(n+1)}{2}.$$

Hence

$$1 + 2 + 3 + \ldots + n + (n+1) = \frac{n(n+1)}{2} + (n+1)$$

$$= (n+1)\left(\frac{n}{2} + 1 \right)$$

$$= (n+1)\left(\frac{n+2}{2} \right).$$

So $1 + 2 + 3 + \ldots + (n+1) = \dfrac{(n+1)\left[(n+1)+1\right]}{2}$, and this equality tells us that $(n + 1) \in S$. Hence $n \in S \Rightarrow (n + 1) \in S$.

The principle of mathematical induction then ensures $S = \mathbb{N}$. Consequently,

(a)
$$1 + 2 + 3 + \ldots + n = \frac{n(n+1)}{2}$$

holds for each natural number n.

Gauss at age ten could have used formula (a) to compute $1 + 2 + 3 + \ldots$, $+ 100 = 5{,}050$. But he did not. Using the formula requires only memory. Gauss did something more sophisticated. He reasoned it out.

We *conjectured* that formula (a) held for each $n = 1, 2, 3, \ldots$ because we *observed* its validity for several values of n. Then we were able *to prove* the validity for each natural number by means of the principle of mathematical induction. But we must be careful of such leaps. Mathematics is not science. Mathematical results are established by proof, not by observation. In this case, we produced a proof so that things turned out fine. But the observations by themselves tell us nothing. Consider the following formula:

$$p = n^2 + n + 41.$$

Let's check the formula for values of n between 1 and 39.

$$n = 1 \Rightarrow p = 1^2 + 1 + 41 = 43$$

$$n = 2 \Rightarrow p = 2^2 + 2 + 41 = 47$$

$$n = 3 \Rightarrow p = 3^2 + 3 + 41 = 53$$

$$\ldots$$

$$n = 39 \Rightarrow p = 39^2 + 30 + 41 = 1,601.$$

All the values of p for $n = 1, 2, \ldots, 39$ are listed in figure 26. If you examine these values: 43, 47, 53, ..., 1,601, you will be surprised to discover that they all turn out to be prime numbers. (The smaller numbers— 43, 47, ..., 97—are obviously prime. For the others, use the sieve of Erathosthenes.) These facts—and the observation method—lead to the conjecture: *The number* p *given by*

$$p = n^2 + n + 41$$

is prime for each n $\in \mathbb{N}$.

n	p	n	p	n	p
1	43	14	251	27	797
2	47	15	281	28	853
3	53	16	313	29	911
4	61	17	347	30	971
5	71	18	383	31	1033
6	83	19	421	32	1097
7	97	20	461	33	1163
8	113	21	503	34	1231
9	131	22	547	35	1301
10	151	23	593	36	1373
11	173	24	641	37	1447
12	197	25	691	38	1523
13	223	26	743	39	1601

Figure 26: $p = n^2 + n + 41$

Seems reasonable enough—figure 26 shows the conjecture holds for $n = 1, 2, \ldots, 39$. However, mathematical truth depends on *proof* and *not on evidence*. The conjecture is *false*. To see this, let $n = 40$. Then

$$\begin{aligned}
p &= 40^2 + 40 + 41 \\
&= 40(40 + 1) + 41 \\
&= 40 \cdot 41 + 41 \\
&= 41 \cdot (40 + 1) \\
&= 41 \cdot 41 \\
&= 41^2
\end{aligned}$$

which is *not prime*.

Even worse—but less interesting—observational errors are possible. Consider the number r defined by:

$$r = (n-1)(n-2)(n-3) \ldots (n-10{,}000).$$

Obviously, $r = 0$ for $n = 1, 2, 3, \ldots, 10{,}000$. (In each case, one factor is zero.) This provides a wealth of observational evidence and a practical person might conclude that $r = 0$ for each $n \in \mathbb{N}$. But neither mathematicians nor poets are practical people. We simply set $n = 10{,}001$ and notice that

$$\begin{aligned}
r &= (10{,}001 - 1)(10{,}001 - 2) \ldots (10{,}001 - 10{,}000) \\
&\neq 0
\end{aligned}$$

because none of the factors equals zero. So, the practical conclusion is *false*.

THE WELL-ORDERING PRINCIPLE

The natural numbers contain a smallest element and its name is 1. If n is any natural number other than 1, then $n > 1$. The analogous statement for the integers does not hold. Since $0 > -1 > -2 > , \ldots,$ $\mathbb{Z}$ does not contain a smallest element. Similarly, $\mathbb{Q}$ does not possess a smallest member.

Considerations of this type become more interesting—and more important—when we consider subsets. Consider the subset of the rational numbers

$$S = \{q : q \in \mathbb{Q}, \ 1 < q\}$$

S, then, contains exactly those rational numbers that exceed 1. The number 1 is a *lower bound* for S, that is, each member of S is larger than 1. But S contains no smallest element.

To see this, let q be an arbitrary element of S. Then $q \in \mathbb{Q}$ and $q > 1$. Let

$$r = \frac{1+q}{2}.$$

Then r is a rational number. (Why?) Moreover, being the average value of 1 and q

$$1 < r < q.$$

This tells us two things: $r \in S$ and $r < q$. Consequently, for any given member of S (namely, q) there is a *smaller* member of S (namely, r). Therefore, S *contains no smallest element.*

Thus not every nonempty subset of the rational numbers contains a smallest element, not even—as S shows—all those subsets that have lower bounds. (Of course, many such subsets of do contain smallest elements.

Examples are $R = \{q : q \in \mathbb{Q} \text{ and } 1 \le q\}$ and $T = \left\{\dfrac{1}{2}, \dfrac{1}{3}, \dfrac{1}{4}\right\}$.)

A nicer situation prevails for the natural numbers. More precisely, we have the *well-ordering principle for the natural numbers:*

Let $S \subset \mathbb{N}$ and $S \ne \varnothing$. Then there exists $m \in S$ such that

$$n \in S \Rightarrow m \le n.$$

Here are some comments:

- The principle asserts that each nonempty subset of the natural numbers contains a smallest element.
- Each subset S of $\mathbb{N}$ has a lower bound; that is, $n \in S \Rightarrow 1 \le n$, so no lower-bound hypothesis need be stated.

- The hypothesis $S \neq \emptyset$ is necessary since the empty set—like Archibald MacLeish's end-of-the-world circus tent[3]—contains nothing, nothing, nothing at all. No way $\emptyset$ can contain a smallest element.
- You can convince yourself of the obviousness of the principle by constructing various nonempty subsets of $\mathbb{N}$. Try as you may, you will not build one that contains no smallest member.

The preceding comment asserts the existence of convincing evidence for the validity of the well-ordering principle for the natural numbers. But we know that evidence will not suffice. Mathematical truth requires proof. The standard proof uses the principle of mathematical induction. Since this proof is more sophisticated than others we have examined, I have chosen to omit it.

The well-ordering principle for the natural numbers extends easily to subsets of the nonnegative integers. That is, *if $\mathcal{U}$ is a nonempty subset of $\{0, 1, 2, 3, \ldots\}$ then $\mathcal{U}$ contains a smallest element.* (If $0 \in \mathcal{U}$ then 0 is the smallest element. Otherwise $\mathcal{U} \subset \mathbb{N}$ and so has a smallest member by the well-ordering principle.) We will make use of this extension when we discuss the theorem that justifies the process of long division.

DIVISION ALGORITHM

In schools students learn by rote a process called "long division." Through imitation and repetition they develop a technique for "dividing" a given integer by another integer called the "divisor." The process—which consists mainly of trial and error—terminates in a finite number of steps and produces two more integers: a "quotient" and a "remainder."

A process of this type—one that produces a solution in a finite number of steps—mathematicians call an *algorithm*. Schoolchildren work at the division algorithm the way a computer works at writing poetry—mechanically, dully, and without understanding. What the children do with division is neither pleasant nor instructive. And more students have been turned off mathematics by long division than have been repelled from literature by the forced reading of *Silas Marner*.

Perhaps students need to learn long division (and maybe Silas *Marner*). But—from a mathematician's point of view (and from ours)—it is more important to know that quotients and remainders exist. Their existence is guaranteed by a theorem called:

The division algorithm: Let a *and* d *be integers with* d > 0. *Then there exist unique integers* q *and* r *with* $0 \leq r < d$ *such that*

$$a = qd + r.$$

(The integers q and r are called, respectively, the *quotient* and the *remainder* in the process of division of a by the divisor d.)

An intuitive argument (which falls short of a proof) convinces us quickly that q and r exist.

Consider the standard ordering of the integers:

$$\ldots < -3 < -2 < -1 < 0 < 1 < 2 < 3 < \ldots.$$

Multiply throughout by the positive integer d. This gives

$$\ldots < -3d < -2d < -d < 0 < d < 2d < 3d < \ldots.$$

Since a is an integer either it equals one of these multiples of d, say $a = qd$, or else it falls between some qd and $(q + 1)d$; that is, $qd < a < (q + 1)d$.

In the first case $a = qd + 0$; that is, $r = 0$. In the second case $a = qd + r$ with

(i) $r > 0$ because $a > qd$ and

(ii) $r < d$ because $a < (q + 1)d = qd + d$.

Together, the two cases give

$$a = qd + r, 0 \leq r < d.$$

A more rigorous proof requires the well-ordering principle for the natural numbers. The proof is tedious and is omitted as is the easier proof that the numbers q and r are unique.

The division algorithm guarantees, for example, that there exists unique integers q and r with $0 \leq r < 14$ such that

$$920 = q(14) + r.$$

In schools you learn to find q and r like this:

$$
\begin{array}{r}
65 \\
14\overline{)920} \\
\underline{84} \\
80 \\
\underline{70} \\
10
\end{array}
$$

Hence,

$$920 = (65)(14) + 10.$$

GREATEST COMMON DIVISOR

Schoolchildren also encounter a division-related concept known as the *greatest common divisor*. This notion associates a number with a given *pair* of positive integers. For example, consider the positive integers 8 and 36. The positive divisors of 8 are 1, 2, 4, 8. The positive divisors of 36 are 1, 2, 3, 4, 6, 9, 12, 18, 36. The numbers common to these two lists are 1, 2, 4. The largest of these, the number 4, is called the *greatest common divisor* of 8 and 36. We write

$$gcd\,(8,\ 36) = 4.$$

More precisely we have:

DEFINITION 3. Let a and b be integers that are not both zero. The positive integer d is called the greatest common divisor of a and b and we write

$$d = gcd\,(a,\ b)$$

if and only if

(i) $d \mid a$ and $d \mid b$

(ii) $m \mid a$ and $m \mid b \Rightarrow m \leq d$.

The definition excludes the case $a = 0$, $b = 0$ because $m \mid 0$ for each $m \in \mathbb{Z}$ so the notion of greatest divisor becomes meaningless. Although the definition allows a and b to be negative, it requires d to be positive. Thus, you need not consider the negative divisors of *a* or *b* in the determination of *d* = *gcd* (*a, b*). For example, the divisors of 8 are 1, –1, 2, –2, 4, –4, 8, –8 while those of 36 are 1,–1, 2, –2, 3, –3, 4, –4, 6, –6, 9, –9, 12, –12, 18, –18, 36, –36. If we let the symbol ±*n* denote either *n* or negative *n*, then these lists may be written as ±1, ±2, ±4, ±8 and ±l, ±2, ±3, ±4, ±6, ±9, ±12, ±18, ±36, respectively. The common divisors are ±1, ±2, ±4. The largest of these—the number 4—remains unchanged from what we found when we looked only at the positive divisors of 8 and 36.

The following theorem establishes the existence of *d* = *gcd* (*a, b*) and, simultaneously, shows that *d* may be written as a *linear combination of* a *and* b. The proof turns on the well-ordering principle for the natural numbers and illustrates once again the fundamental role played by this result and its companion, the principle of mathematical induction. Because of its sophistication I will not give the proof here.

THEOREM 2. Let *a* and *b* be integers that are not both zero. Then *d* = *gcd* (*a, b*) exists and

$$d = ma + nb$$

for some integers *m* and *n*.

Before we proceed further with these ideas, I should point out that our definition of greatest common divisor is not quite standard. Definition 2 essentially says:

$$d = gcd(a,b) \Leftrightarrow \begin{cases} d > 0 \\ d \mid a \text{ and } d \mid b, \text{ and} \\ m \mid a \text{ and } m \mid b \Rightarrow m \leq d. \end{cases}$$

In the standard definition the third line on the right-hand side of the double implication is replaced with

$$m \mid a \text{ and } m \mid b \Rightarrow m \mid d.$$

Mathematicians usually prefer the standard definition because it does not involve the concept of order and thus may be extended to more general settings that allow the concept of "divisor" but not that of "greater than." I prefer definition 1 precisely because it contains the order notion that one expects to be reflected in the phrase "greatest common denominator." But the choice is a matter of taste. In the case of the integers, they are equivalent. (We need not bother with the proof.)

Theorem 2 provides an example of what mathematicians call an *existence theorem*. The result ensures the existence of $d = gcd\,(a, b)$ and asserts that it may be written as a linear combination of a and b. However, theorem 2 does not tell us how to find the greatest common divisor. For this we need a theorem of the type that mathematicians call *constructive*. We need a step-by-step procedure for the determination of $d = gcd\,(a, b)$.

An ancient procedure of this type bears the name *Euclidean algorithm* (even though it may predate Euclid). The basic idea of the Euclidean algorithm is this:

> *Given two positive integers, divide the smaller into the larger and obtain a quotient and a remainder. If the remainder is positive, divide it into the previous divisor. Continue until you obtain a zero remainder. Stop. The last nonzero remainder is the greatest common divisor of the two given integers.*

The proof that this procedure produces the greatest common divisor requires only respective use of the division algorithm. It is, however, tedious to describe and I will omit it.

The following illustrates the procedure:

EXAMPLE. Find the greatest common divisor of 102 and 16.

The process may be systemized by:

$$
\begin{array}{r}
6 \\
16\overline{)102} \\
\underline{96} \\
\end{array}
$$

$$
\begin{array}{r}
2 \\
6\overline{)16} \\
\underline{12} \\
\end{array}
$$

$$
\begin{array}{r}
1 \\
4\overline{)6} \\
4 \\
\end{array}
$$

$$
\begin{array}{r}
2 \\
2\overline{)4} \\
4 \\
\hline
0. \\
\end{array}
$$

The last nonzero remainder is $r_3 = 2$. Thus, $2 = gcd$ (102, 16).

The corresponding equations are:

$$102 = 6(16) + 6$$
$$16 = 6(2) + 4$$
$$6 = 4(1) + 2$$
$$4 = 2(2) + 0.$$

Proceeding upward from the next-to-last equation gives:

$$2 = 6 - 4(1) = 6 - 4$$
$$= 6 - [16 - 6(2)] = 3(6) - 16$$
$$= 3[102 - 6(16)7] - 16$$
$$= 3(102) - 19(16).$$

Therefore,

$$2 = gcd \ (102, 16) = 3(102) + (-19)(16)$$

as promised by theorem 2.

Since gcd $(a, b) = gcd$ $(-a, -b)$, nothing has been lost by the restriction to positive integers in the Euclidean algorithm discussion. For example, gcd $(-102, 16) = gcd$ $(102, 16) = 2$.

Of course, the greatest common divisor of two integers may turn out to be 1.

EXAMPLE. Find $d = gcd$ (112, 15).
SOLUTION. The Euclidean algorithm gives

$$\begin{array}{r} 7 \\ 15\overline{)112} \\ \underline{105} \end{array}$$

$$\begin{array}{r} 2 \\ 7\overline{)15} \\ \underline{14} \end{array}$$

$$\begin{array}{r} 7 \\ 1\overline{)7} \\ \underline{7} \\ 0. \end{array}$$

Thus, $d = gcd\,(112, 15) = 1$. Here, the corresponding equations are

$$112 = 7 \cdot (15) + 7$$
$$15 = 7(2) + 1$$
$$2 = 1(2).$$

These give:

$$1 = 15 - 7(2)$$
$$1 = 15 - [112 - 7(15)](2)$$
$$1 = 15(15) - (2)(112)$$
$$1 = 15(15) + (-2)(112)$$

and the last expression is of the form $1 = M(15) + N(112)$.

When this occurs, the given integers are said to be *relatively prime*.

DEFINITION 4. Let a and b be integers that are not both zero. We say that a and b are relatively prime if and only if $gcd\,(a, b) = 1$.

Thus 112 and 15 are relatively prime. (But neither number is prime.)

If a and b are relatively prime then $1 = gcd\,(a, b)$ so by theorem 2, $1 = ma + nb$, for some integers m and n.

On the other hand, suppose that $1 = ma + nb$ for some integers m and n. Let $d = gcd\,(a, b)$. Then $d|a$ and $d|b$. It follows from parts (i) and (ii) of theorem 1 that $d(ma + nb)$. Thus $d|1$. Since d is a positive integer this means $d = 1$.

These remarks constitute the proof of

THEOREM 3. Let a and b be integers that are not both zero. Then $gcd\,(a, b) = 1$ (i.e., a and b are relatively prime) if and only if

$$1 = ma + nb$$

for some integers m and n.

A result that follows easily from a given theorem is often called a *corollary*. An example is:

COROLLARY 1. Let a, b, $s \in \mathbb{Z}$, $a \neq 0$. Suppose $gcd\,(a, b) = 1$. If $a|sb$ then $a|s$.

PROOF. By theorem 3

$$1 = ma + nb$$

for some integers m and n. So

$$s = s\,(ma + bn)$$

or

$$s = sma + sbn.$$

Since $a|sb$ it follows that $a|sbn$ (theorem 1 (ii)). Also $a|sma$. Thus $a|\,(sma + sbn)$ (theorem 1(i)). Hence $a|s$.

An elementary recasting of this result gives:

COROLLARY 2. Suppose p is prime and $p|ab$. Then $p|a$ or $p|b$.

PROOF. If $p|b$ the corollary holds. So suppose $p \nmid b$. Then $gcd\,(p, b) = 1$. (The only divisors of p are p, $-p$, 1, -1.) Then corollary 1 gives $p|b$.

As an example of corollary 2 notice that $189 = 7(27)$. Hence, $7|189$. But also $189 = 9(21)$. Thus $7|9(21)$. Corollary 2 ensures that $7|9$ or $7|21$. (Obviously, $7|21$.)

Notice that corollary 2 fails if p is not a prime number. For example, $16|8\,(20)$ but $16 \nmid 8$ and $16 \nmid 20$.

Corollary 2 may be extended to a result involving the product of three or more integers. For example, suppose $p|(ab)c$ and p is a prime. Thus by corollary 2, $p|(ab)$ or $p|c$. If $p \nmid c$ then we must have $p|ab$. Using corollary 2 again we obtain $p|a$ or $p|b$. Combining all this gives:

$$p \text{ a prime and } p|abd \Rightarrow p|a \text{ or } p|b \text{ or } p|c.$$

The preceding argument may be extended in a natural way to products of more than three integers. The proof is left as an exercise.

COROLLARY 3. Let $a_1, a_2, a_3, \ldots, a_n$ be nonzero integers and n a natural number. Then, if p is prime,

$$p \mid a_1 a_2 a_3 \ldots a_n \Rightarrow p \mid a_1 \text{ or } p \mid a_2 \ldots \text{ or } p \mid a_n.$$

The notion of *congruence* goes back at least as far as Gauss and the *Disquisitiones Aritmeticae.*

DEFINITION 5. Let a, b, and m be integers with $m > 0$. We say that a and b are *congruent modulo m* and we write

$$a \equiv b \bmod m$$

if and only if

$$m \mid (a - b).$$

That is, if and only if

$$a - b = km, \ k \in \mathbb{Z}.$$

For example,

$$13 \equiv 8 \bmod 5 \text{ because } 13 - 8 = 1 \cdot 5$$
$$26 \equiv 4 \bmod 2 \text{ because } 26 - 4 = (11)2$$
$$60 \equiv 88 \bmod 7 \text{ because } 60 - 88 = (-4)7.$$

Because any two integers are congruent modulo 1 $(a - b = (a - b) \cdot 1)$, we often require $m \neq 1$ when we speak of congruence modulo m. Notice that definition 1 requires $m > 0$, so we usually deal only with $m > 1$.

Some basic properties of congruence flow easily from the definition. In particular congruences may be added and multiplied like ordinary equations.

THEOREM 5. Let a, b, c, d, m be integers with $m \geq 1$. Then

(i) $a \equiv b \bmod m$ and $c \equiv d \bmod m \Rightarrow (a + c) \equiv (b + d) \bmod m$

and

(ii) $a \equiv b \bmod m$ and $c \equiv d \bmod m \Rightarrow ac \equiv bd \bmod m.$

PROOF.

(i)

$$a \equiv b \bmod m$$

$$c \equiv d \bmod m \Rightarrow m \mid (a-b) \text{ and } m \mid (c-d)$$

$$\Rightarrow a-b = k_1 m, \ c-d = k_2 m \text{ for some } k_1 k_2 \in \mathbb{Z}$$

$$\Rightarrow (a-b)+(c-d) = k_1 m + k_2 m$$

$$\Rightarrow (a+c)+[(-b)+(-d)] = (k_1 + k_2)m$$

$$\Rightarrow (a+c)-(b+d) = (k_1 + k_2)m$$

$$\Rightarrow m \mid [(a+c)-(b+d)]$$

$$\Rightarrow a+c \equiv (b+d) \bmod m.$$

(ii) As in (i)

$$a \equiv b \bmod m, \ c \equiv d \bmod m \Rightarrow a-b = k_1 m$$

$$c-d = k_2 m$$

$$\text{for some } k_1, k_2 \in \mathbb{Z}.$$

Hence, $a = k_1 m + b$ and $c = k_2 m + d$. Thus,

$$ac = (k_1 m + b)(k_2 m + d)$$

$$= k_1 k_2 m^2 + k_1 md + bk_2 m + bd.$$

So

$$ac - bd = (k_1 k_2 m + k_1 d + bk_2)m$$

and then

$$m \mid (ac - bd).$$

Therefore

$$ac \equiv bd \bmod m.$$

For example, $10 \equiv 2 \bmod 4$ and $15 \equiv 3 \bmod 4$. ($4|(10-2)$ and $4|(15-3)$.) Hence, by theorem 5,

$$10 + 15 \equiv (2+3) \bmod 4$$

and

$$(10)(15) \equiv 2 \cdot 3 \bmod 4.$$

(These congruences may be easily verified directly since they assert, respectively, $25 \equiv 5 \bmod 4$ and $150 \equiv 6 \bmod 4$.)

Notice that, since $r \equiv r \bmod m$, we have as an easy consequence:

COROLLARY 4. $a \equiv b \bmod m \Rightarrow a + r \equiv (b + r) \bmod m$, $ar \equiv br \bmod m$, for any integer r.

This means that any integer may be added to both sides of a congruence and a congruence may be multiplied by any given integer. For example, $10 \equiv 2 \bmod 4$ so $10 + r \equiv (2 + r) \bmod 4$ and $10r \equiv 2r \bmod 4$ hold for any integer r.

If we take $c = a$ and $d = b$ in part (ii) of theorem 5, we obtain

$$a \equiv b \bmod m \Rightarrow a \cdot a \equiv b \cdot b \bmod m.$$

Thus

$$a \equiv b \bmod m \Rightarrow a^2 \equiv b^2 \bmod m.$$

Repeating the argument gives

$$a \equiv b \bmod m \Rightarrow a^3 \equiv b^3 \bmod m.$$

These observations lead us to:

COROLLARY 5. $a \equiv b \bmod m$ and $n \in \mathbb{N} \Rightarrow a^n \equiv b^n \bmod m$, for $a, b, \in \mathbb{Z}$ with $a \equiv b \bmod m$.

The proof is by induction on n and is omitted. The corollary says, for example, $10 \equiv 2 \bmod 4$. So, $10^6 \equiv 2^6 \bmod 4$ or $1{,}000{,}000 = 64 \bmod 4$.

The division algorithm plays a significant role in the collection of ideas associated with the congruence notion. To see this, let a, m be integers with $m > 0$.

The division algorithm says

$$a = mq + r, \ 0 \le r < m.$$

Hence

$$a - r = mq$$

so

$$a \equiv r \bmod m.$$

This tells us that *any integer is congruent* modulo m *to the remainder obtained when* it is divided by *m*. But the possible remainders are only the integers 0, 1, 2, 3, . . . , *m* − 1. These remainders provide the proof of:

THEOREM 6. Let *a* and *m* be integers with *m* > 0. Then *a* is congruent *modulo m* to exactly one of the integers 0, 1, 2, 3, . . . , *m* − 1.

For example, if *m* = 4 then the possible remainders are 0, 1, 2, 3. Consequently, if $a \in \mathbb{Z}$, then exactly one of the following hold:

$$a \equiv 0 \bmod 4$$
$$a \equiv 1 \bmod 4$$
$$a \equiv 2 \bmod 4$$
$$a \equiv 3 \bmod 4.$$

Theorem 6 may be extended to give

THEOREM 7. Let *a*, *b*, *m* be integers with *m* > 0. Then

$a \equiv b \bmod m \Leftrightarrow a$ and *b* have the same remainder when divided by *m*.

PROOF. Suppose *a* and b have the same remainder when divided by *m*. Then

$$a = q_1 m + r$$

and

$$b = q_2 m + r$$

for some $q_1,\, q_2,\, r \in \mathbb{Z}$ and $0 \le r < m$. Then

$$a - b = (q_1 m + r) - (q_2 m + r)$$
$$= q_1 m - q_2 m$$
$$= (q_1 - q_2)m.$$

Hence

$$a \equiv b \bmod m.$$

This proves the right-to-left half of the above double implication. The proof in the other direction remains as an exercise.

Theorem 5 suggests that *congruence modulo m* behaves much like ordinary equality. The similarity can be deepened if we isolate the fundamental properties that belong to the concept of equality. These properties bear the names *reflexive*, *symmetric*, and *transitive*, and are, respectively:

$$a = b \qquad \text{(reflexive property)}$$

$$a = b \Rightarrow b = a \qquad \text{(symmetric property)}$$

$$a = b, \ b = c \Rightarrow a = c \quad \text{(transitive property).}$$

The following easy result shows that these properties are inherited by the congruence notion:

THEOREM 8. Let a, b, c, m be integers with $m > 0$. Then

$$\text{(i)} \ a \equiv a \bmod m \qquad \textit{(reflexive)}$$
$$\text{(ii)} \ a \equiv b \bmod m \Rightarrow b \equiv a \bmod m \qquad \textit{(symmetric)}$$
$$\text{(iii)} \ a \equiv b \bmod m, \text{ and } b \equiv c \bmod m \Rightarrow a \equiv c \bmod m \qquad \textit{(transitive).}$$

PROOF.   (i) $a - a = 0 = m \cdot 0$, so $a \equiv a \bmod m$.

(ii)

$$a \equiv b \bmod m \Rightarrow m \mid (a - b)$$
$$\Rightarrow m \mid (b - a)$$
$$\Rightarrow b \equiv a \bmod m.$$

(iii)

$$a \equiv b \bmod m, \ b \equiv c \bmod m \Rightarrow m \mid (a - b) \text{ and } m \mid (b - c)$$
$$\Rightarrow m \mid [(a - b) + (b - c)] \quad \text{(theorem 1(i))}$$
$$\Rightarrow m \mid (a - c)$$
$$\Rightarrow a \equiv c \bmod m.$$

Gauss himself chose the "$\equiv$" symbol for congruence. And he chose it precisely because of the similarities between the notion of congruence and the basic concept of equality. Theorem 5 and theorem 8 describe much of this similarity.

Actually, both equality and congruence are special cases of something more general. This "something" is called an *equivalence relation.*

EQUIVALENCE RELATIONS

We may think of ordinary equality as an example of a *relation.* For example, we might say: "two integers m and n are related if and only if $m = n$." Similarly congruent modulo m provides another relation between pairs of integers. We would say: "a and b are related if and only if $a \equiv b \bmod m$."

These two relations are different (except for the case $m = 1$) and we have a unique symbol for each: "=" for equality and "$\equiv$" for congruence. Moreover, many other relations are possible. An example is the relation given by "greater than": "a is related to b if and only if $a > b$." This defines yet another relation on the integers (also on the rationals) and we denote it by the third symbol ">."

If we want to examine the general notion of relation we will need some general symbol of identification. Several symbols appear in the mathematical literature and to begin we will select one of those most commonly used. We will write $a \sim b$ to indicate that a and b are related by some fixed—but perhaps unknown—rule of association. (Many mathematicians read the symbol $a \sim b$ as "a tilde b.")

In general, *a relation $\sim$ is defined on pairs of elements of a given set S.* The key idea is that, given $a, b \in S$ either $a \sim b$ holds or it does not hold. (In the latter case we may write $a \nsim b$.) However, we often abuse the language somewhat and say:

let $\sim$ be a relation on S

rather than

let $\sim$ be a relation on the pairs of elements of S.

Occasionally, we may illustrate the relation concept by real-world examples. For instance, let P denote the set of all living people. Define the relation $\sim$ on P by

$p \sim q \Leftrightarrow p$ and q have the same biological mother.

This rule appropriately defines a relation on P since any two people either have the same mother or they do not.

Those particular relations that share certain fundamental properties are called *equivalence relations*.

DEFINITION 6. Let $\sim$ be a relation on a nonempty set S. We say that $\sim$ is an equivalence relation if and only if for each a, b, c in S:

> (i) $a \sim a$ ($\sim$ is reflective)
>
> (ii) $a \sim b \Rightarrow b \sim a$ ($\sim$ is symmetric)
>
> (iii) $a \sim b,\ b \sim c \Rightarrow a \sim c$ ($\sim$ is transitive).

For example:

- Equality is an equivalence relation on each nonempty set by its very definition.
- Congruence modulo m is an equivalence relation on $\mathbb{Z}$ by theorem 8.
- "Greater than" is not an equivalence relation on $\mathbb{Z}$ since it is neither reflexive nor symmetric. ($a > a$ never holds and $a > b$ does not imply $b > a$. This relation is, however, transitive.)
- "Is the mother of" is not an equivalence relation on the set of all people. (In particular, it is not reflexive: "p is the mother of p" fails to hold.)

An equivalence relation neatly partitions the given set into subsets called *equivalence classes*.

DEFINITION 7. Let $\sim$ be an equivalence relation on a set S. For each $x \in S$, let

$$C(x) = \{y : y \in S,\ y \sim x\}.$$

The set $C(x)$ is called the equivalence class determined by x.

Thus the equivalence class $C(x)$ consists of all those elements of S that are related to x by the equivalence relation $\sim$. As an example, let $S = \mathbb{Z}$ and let $\sim$ be defined on $\mathbb{Z}$ by

$$x \sim y \Leftrightarrow x \equiv y \bmod m$$

where $x, y \in \mathbb{Z}$ and m is a fixed nonzero integer. We know that $\sim$ so defined is an equivalence relation (theorem 8). Here

$$C(x) = \{y : y \equiv x \bmod m\}$$

is just the set of integers congruent modulo m to the integer x.

As a particular case, consider $m = 3$. Then $C(x) = \{y : y \equiv x \bmod 3\}$. At first glance there appear to be infinitely many of these equivalence classes, one for each $x \in \mathbb{Z}$. But such is not the case. To see this look at $C(0)$, $C(1)$, $C(2)$. Then

$$C(0) = \{y : y \equiv 0 \bmod 3\} = \{\ldots, -6, -3, 0, 3, 6, \ldots\}$$
$$C(1) = \{y : y \equiv 1 \bmod 3\} = \{\ldots, -5, -2, 1, 4, 7, \ldots\}$$

and

$$C(2) = \{y : y \equiv 2 \bmod 3\} = \{\ldots, -4, -1, 2, 5, 8, \ldots\}.$$

Note that the sets $C(0)$, $C(1)$, and $C(2)$ may be written as

$$C(0) = \{\ldots, -9, -6, -3, 0, 3, 6, 9, \ldots\}$$
$$C(1) = \{\ldots, -8, -5, -2, 1, 4, 7, 10, \ldots\}$$
$$C(2) = \{\ldots, -7, -4, -1, 2, 5, 8, 11, \ldots\}$$

respectively. Each integer is contained in one of these sets so that

$$\mathbb{Z} = C(0) \cup C(1) \cup C(2).$$

Moreover, each integer belongs to exactly one of these sets. Thus, $C(0) \cap C(1) = \emptyset$, $C(0) \cap C(2) = \emptyset$, $C(1) \cap C(2) = \emptyset$. This says that any distinct pair of the sets $C(0)$, $C(1)$, and $C(2)$ are disjoint. Mathematicians describe this by saying simply: "The sets $C(0)$, $C(1)$, and $C(2)$ are pairwise disjoint."

The apparent infinity of equivalence classes,

$$\ldots, C(-2), C(-1), C(0), C(1), C(2), \ldots$$

telescopes down to just these three:

$$C(0), C(1), C(2)$$

which are exactly the equivalence classes determined by the possible remainders obtained upon division by 3. Theorem 7 took care of the others. For instance, $C(33) = C(0)$, because 33 and 0 have the same remainder, namely, 0 when divided by 3. Similarly, $C(26) = C(2)$, $C(19) = C(1)$, and so on.

The above argument dealt with the case of congruence modulo 3. The argument extends in a natural way to the general case of congruence modulo m where m is any positive integer. (The necessary notation is cumbersome, so details are omitted.)

The result is that, in the case of congruence modulo m, the equivalence classes become

$$C(0), \; C(1), \; C(2), \; \ldots \; C(m-1)$$

and, once again,

$$\mathbb{Z} = C(0) \cup C(2) \cup \ldots \cup C(m-1)$$

and these equivalence classes are pairwise disjoint. And, in this setting, the equivalence classes are given a special name.

DEFINITION 8. Let m be a positive integer. Let $\sim$ be the equivalence relation on $\mathbb{Z}$ given by

$$x \sim y \Leftrightarrow x \equiv y \bmod m.$$

Let

$$C(x) = \left\{ y : y \equiv x \bmod m \right\}$$

be the equivalence class determined by x. The equivalence classes

$$C(0), \; C(1), \; C(2), \; \ldots, \; C(m-1)$$

are called the residue classes modulo m.

The preceding discussion shows that these m residue classes compose the entire collection of equivalence classes and form a pairwise disjoint partition of $\mathbb{Z}$. Because of the importance of this particular example, it is convenient to have a special symbol for the residue classes. A slightly defective symbol that appears frequently in the literature is:

$$[n] = \{\kappa : \kappa \in \mathbb{Z}, \; \kappa \equiv n \bmod m\}.$$

(The defect lies in the absence of the integer m in the symbol $[n]$. Since it does not appear, we must understand from the context that we are dealing with congruence modulo the fixed integer m. A better symbol—not in common use—would be $[n]_m$. The general symbol for an equivalence class, $C(x)$, contains a similar defect, since it does not show the equivalence relation $\sim$.)

With this new notation—and the integer $m > 0$ given in advance—we have

$$\mathbb{Z} = [0] \cup [1] \cup [2] \cup \ldots \cup [m-1]$$

when the residue classes shown are pairwise disjoint.

Let us now return to the general setting of an arbitrary nonempty set S, an equivalence relation $\sim$ on S and equivalence classes $C(x) = \{y : y \in S, y \sim x\}$. The following theorem shows that—even in the general setting—the equivalence classes partition S into pairwise disjoint sets. Prior to the theorem, we need a preliminary result (called a lemma):

LEMMA 1. Let $\sim$ be an equivalence relation on a nonempty set S. Let x, $y \in S$, and let $C(x)$ and $C(y)$ be the equivalence classes determined, respectively, by x and y. Then either

$$C(x) = C(y)$$

or

$$C(x) \cap C(y) = \emptyset.$$

(Thus any two equivalence classes are either identical or else they contain no element in common.)

PROOF. If $C(x) \cap C(y) = \emptyset$ the proof is complete. Suppose $C(x) \cap C(y) \neq \emptyset$. Then there exists $z \in S$ such that $z \in C(x) \cap C(y)$; that is, $z \in C(x)$ and $z \in C(y)$. But

$$z \in C(x) \Rightarrow z \sim x \text{ (definition of } C(x))$$

and

$$z \in C(y) \Rightarrow z \sim y \text{ (definition of } C(y)).$$

Then

$$z \sim x \text{ and } z \sim y \Rightarrow x \sim z \text{ and } z \sim y \text{ ($\sim$ is symmetric)}$$

$$\Rightarrow x \sim y \text{ ($\sim$ is transitive)}.$$

Now let $z \in C(x)$. Then $z \sim x$, by definition of $C(x)$. But, from the above, $x \sim y$. Hence, $z \sim y$, because $\sim$ is transitive.) Then

$$z \in C(x) \Rightarrow z \in C(y).$$

But this means

$$C(x) \subset C(y).$$

A similar argument gives $C(y) \subset C(x)$. (Start with $z \in C(y)$ and repeat the steps.) Hence,

$$C(x) = C(y)$$

and the proof is complete.

In our earlier example of the integers and the equivalence relation determined by congruence modulo m, the equivalence classes turned out to be the residue classes: $[0], [1], \ldots, [m-1]$. Since there are only m of them, we can indicate their union—which turns out to be $\mathbb{Z}$—by

$$\mathbb{Z} = [0] \cup [1] \cup \ldots [m-1].$$

But, in the case of an arbitrary set S and an unknown equivalence relation $\sim$, we cannot be certain that there exist only finitely many equivalence classes $C(x)$. Therefore, we need a new symbol to indicate the possibly infinite union of all of them. One such symbol is $\cup_{x \in S} C(x)$, which means precisely:

$$\cup_{x \in S} C(x) = \{y : y \in C(x) \text{ for some } x \in S\}.$$

The promised theorem guarantees that this union fills S.

THEOREM 9. Let $\sim$ be an equivalence relation of a nonempty set S. Let $C(x)$ denote the equivalence class corresponding to $x \in S$. Then

$$S = \cup_{x \in S} C(x)$$

and any two equivalence classes are either identical or disjoint.

PROOF. Because of lemma 1, we must only prove that

$$S = \cup_{x \in S} C(x).$$

Since each $C(x)$ contains only members of S, clearly

$$\cup_{x \in S} C(x) \subset S.$$

Now let $w \in S$. Then $w \sim w$ because $\sim$ is reflexive. Thus, $w \in C(w)$, which is one of the sets in given union. Hence

$$w \in \cup_{x \in S} C(x).$$

Thus,

$$S \subset \cup_{x \in S} C(x).$$

Therefore

$$S = \cup_{x \in S} C(x)$$

and the proof is complete.

Figure 27 illustrates this theorem by showing S as a rectangle and the equivalence classes as diagonal pieces. The three dots in the sketch indicate the possibility of infinitely many such pieces.

As a real-world example, consider a town with the property that each of its citizens lives in a house. Let P be the set of all people in the town. Define on p by

$$p \sim q \Leftrightarrow \text{"}p \text{ and } q \text{ live in the same house."}$$

It is easy to see that $\sim$ is reflexive, symmetric, and transitive and, therefore, is an equivalence relation. For each p, the equivalence class $C(p)$ consists of those persons who live in the same house as does p. Here the shaded regions of figure 27 are determined by the houses in the town and each consists physically of the people living in any one of them.

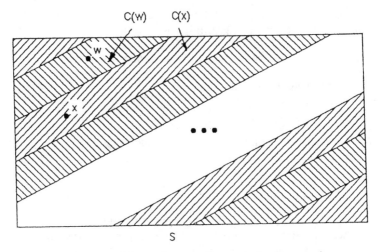

Figure 27: Partition of a set into equivalence classes

Incidentally, theorem 9 has a valid converse which asserts that *any* partition of a set into such a union of sets induces an equivalence relation onto the set. Time spent struggling with the exact statement of the converse and its proof would be well spent.

GROUPS

The subject taught in schools called algebra consists mainly of the presentation of rules governing the proper manipulation of symbols representing numbers. When a child tells you he does not like algebra, he is saying he does not like to manipulate symbols. But this is not what a mathematician means by "algebra." A mathematician considers algebra to be roughly that part of the mathematical world whose ideas and techniques involve abstractions and generalizations of basic properties of familiar numbers, such as the integers or the rationals. A *number theorist* studies the integers. An *algebraist* studies generalizations of properties possessed by the integers.

For example, we know from theorem 7 of chapter 3 that the fundamental properties of integer addition are: the sum of any two integers remains an integer, addition is commutative and associative, and there exists an additive inverse. A direct generalization of these properties to an arbitrary set yields a fundamental—and truly algebraic—object called a group.

DEFINITION 9. Let $\star$ be a binary operation for a set G. (This means: $x, y \in G \Rightarrow x \star y \in G$ and $x \star y$ is unique.) We say that G, with the operation $\star$, is a group, or that $(G, \star)$ is a group if and only if

   (i) $x, y, z \in G \Rightarrow x \star (y \star z) = (x \star y) \star z$ ($\star$ is associative)

   (ii) there exists $e \in G$ such that
         $e \star x = x \star e = x$
         for all $x \in G$ (e is an identity)

   (iii) for each $x \in G$ there exists $x^{-1} \in G$ such that
          $x \star x^{-1} = x^{-1} \star x = e$ (x has an inverse).

   If, in addition

   (iv) $x, y \in G \Rightarrow x \star y = y \star x$ ($\star$ is commutative),
         we say that G is a commutative group or an Abelian group (after Niels Henrik Abel, 1802–1829).

Theorem 7A of chapter 3 shows that the integers form a group with respect to addition, that is, $(\mathbb{Z}, +)$ is a group. Similarly, theorem 16A of chapter 3 tells us that $(\mathbb{Q}, +)$ is a group. The natural numbers do not form a group with respect to addition because, in particular, $\mathbb{N}$ contains no additive identity.

The group notion seems as uncomplicated and innocent as a flower. But the simplicity is deceptive. Serpents lie underneath. The subarea of algebra called *group theory* crawls with deep and difficult problems as a jungle crawls with snakes. These woods are dark and deep. If you want to be a group theorist, you must first of all be brave.

One recently solved problem, for example, involves only groups with finitely many elements that, in addition, possess such nice properties that they are called *simple groups*. Should be easy. But this single problem—the so-called classification problem—required for its solution the work of scores of mathematicians and filled about ten thousand pages of published research papers. Placing man on the moon was a less formidable achievement.

We shall not here pursue the theory of groups. Instead, we will content ourselves with an exercise or two and a single, significant example that illustrates a connection between number theory and algebra.

Consider the integers $\mathbb{Z}$ and a fixed integer $m > 0$. As before let $[\kappa]$ denote the residue class modulo determined by $\kappa \in \mathbb{Z}$. That is,

$$[\kappa] = \{n : n \equiv \kappa \bmod m\}.$$

We know that there exist exactly m of these residue classes: $[0], [1], \ldots,$ $[m-1]$. Let R_m denote the set of all these. Thus,

$$R_m = \{[0], [1], [2], \ldots, [m-1]\}.$$

We might attempt to define a binary operation $\star$ on R_m by

$$[\kappa] \star [\ell] = [\kappa + \ell].$$

Thus the operation $\star$ produces, from the residue class that contains κ and the residue class that contains ℓ, the residue class that contains $\kappa + \ell$. Such an operation will be a binary operation on R_m since $[\kappa + \ell]$ is also a member of R_m *provided it is well defined.*

Let's illustrate the difficulty with an example. Consider the special case $m = 2$. Then

$$R_m = R_3 = \{[0], [1], [2]\}.$$

The above operation gives, in particular,

$$[1] \star [2] = [1 + 2]$$

$$= [3].$$

But [1] and [2] have other representations. For example,

$$[1] = [10] \ (10 \equiv 1 \bmod 3)$$

and

$$[2] = [29] \ (29 \equiv 2 \bmod 3).$$

Hence,

$$[1] \star [2] = [10] \star [29]$$

$$= [39]$$

which seems to be a different answer. Unless the answers agree, the operation $\star$ is not well defined. Fortunately, they turn out to be the same:

$$[39] = [3]$$

because $39 \equiv 3 \bmod 3$ ($39 \in [39]$, $39 \in [3]$ and two equivalence classes are either disjoint or identical). In fact, $[39] = [3] = [0]$ so that

$$[1] \star [2] = [0]$$

which is a member of R_3.

Things work out equally well in the more general setting R_m. Suppose

$$[\kappa] = [\kappa_1] \text{ and } [\ell] = [\ell_1].$$

Then,

$$[\kappa] \star [\ell] = [\kappa + \ell].$$

But also,

$$[\kappa] \star [\ell] = [\kappa_1] \star [\ell_1]$$

$$= [\kappa_1 + \ell_1].$$

Again, the answers agree:

$$[\kappa] = [\kappa_1] \Rightarrow \kappa \equiv \kappa_1 \bmod m$$

$$[\ell] = [\ell_1] \Rightarrow \ell \equiv \ell_1 \bmod m.$$

Hence,

$$\kappa + \ell \equiv (\kappa_1 + \ell_1) \bmod m$$

by theorem 5(i). Hence, $(\kappa + \ell) \in [\kappa_1 + \ell_1]$. But $(\kappa + \ell) \in [\kappa + \ell]$. So

$$[\kappa + \ell] \cap [\kappa_1 + \ell] \neq \emptyset.$$

Thus,

$$[\kappa + \ell] = [\kappa_1 + \ell].$$

This discussion proves:

LEMMA 2. Let $m \in \mathbb{Z}$, $m \neq 0$, and let

$$R_m = \{[0], [1], [2], \ldots, [m-1]\}$$

be the set of residue classes of integers modulo m. Then the operation $\star$ defined by

$$[\kappa] \star [\ell] = [\kappa + \ell]$$

is a binary operation on R_m.

It is convenient—and traditional—to write lemma 2 with different notation. We make two changes:

(a) replace $[\kappa]$ with $\overline{\kappa}$

and

(b) replace $\star$ with +.

The first change introduces yet another symbol for a residue class modulo m. Originally, we used $C(\kappa)$ for the equivalence class that contains κ and then we replaced this by $[\kappa]$, when the given equivalence relation was congruence modulo m. In this case $[\kappa]$ became the residue class determined by m. Now—in order to emphasize the algebraic setting—we are writing $[\kappa] = \overline{\kappa}$.

This change together with the use of + for $\star$ transforms lemma 2 into:

LEMMA 2A. Let $m \in \mathbb{Z}$, $m \neq 0$, and let

$$R_m = \{\overline{0}, \overline{1}, \overline{2}, \ldots, \overline{m-1}\}$$

be the set of residue classes of integers modulo m. Then the operation $+$ defined by

$$\overline{\kappa} + \overline{\ell} = \overline{\kappa + \ell}$$

is a binary operation on R_m.

Once again, we have an example of a single symbol used with different meanings. In

$$\overline{\kappa} + \overline{\ell} = \overline{\kappa + \ell}$$

the $+$ symbol on the left-hand side is the operation being defined. On the right side, $+$ means ordinary integer addition.

If we look once again at the case $m = 3$, we have

$$R_m = R_3 = \left\{ \overline{0}, \overline{1}, \overline{2} \right\}$$

and, in particular,

$$\overline{0} + \overline{1} = \overline{1}$$
$$\overline{1} + \overline{2} = \overline{3} = \overline{0}$$
$$\overline{2} + \overline{2} = \overline{4} = \overline{1}.$$

When $m = 5$,

$$R_m = R_5 = \left\{ \overline{0}, \overline{1}, \overline{2}, \overline{3}, \overline{4} \right\}.$$

Some of the possible sums here are

$$\overline{1} + \overline{3} = \overline{4}$$
$$\overline{2} + \overline{3} = \overline{5} = \overline{0}$$
$$\overline{3} + \overline{4} = \overline{7} = \overline{2}.$$

Figure 28 shows the complete addition table for R_3 and R_5.

With these preliminaries behind, we are ready for our number-theory/group-theory connection.

THEOREM 10. Let $m \in \mathbb{Z}$, $m \neq 0$, and let

$$R_m = \left\{ \overline{0}, \overline{1}, \overline{2}, \ldots, \overline{m-1} \right\}$$

R₃

+	$\bar{0}$	$\bar{1}$	$\bar{2}$
$\bar{0}$	$\bar{0}$	$\bar{1}$	$\bar{2}$
$\bar{1}$	$\bar{1}$	$\bar{2}$	$\bar{0}$
$\bar{2}$	$\bar{2}$	$\bar{0}$	$\bar{1}$

R₅

+	$\bar{0}$	$\bar{1}$	$\bar{2}$	$\bar{3}$	$\bar{4}$
$\bar{0}$	$\bar{0}$	$\bar{1}$	$\bar{2}$	$\bar{3}$	$\bar{4}$
$\bar{1}$	$\bar{1}$	$\bar{2}$	$\bar{3}$	$\bar{4}$	$\bar{0}$
$\bar{2}$	$\bar{2}$	$\bar{3}$	$\bar{4}$	$\bar{0}$	$\bar{1}$
$\bar{3}$	$\bar{3}$	$\bar{4}$	$\bar{0}$	$\bar{1}$	$\bar{2}$
$\bar{4}$	$\bar{4}$	$\bar{0}$	$\bar{1}$	$\bar{2}$	$\bar{3}$

Figure 28: Addition tables for R_3 and R_5

be the set of residue classes of integers modulo m. Let + be defined on R_m by

$$\bar{\kappa} + \bar{\ell} = \overline{\kappa + \ell}$$

then $(R_m, +)$ is a commutative group.

PROOF. Lemma 2 says that + is a binary operation for R_m. We will check, one by one, the other conditions of definition 9.

(i)

$$\bar{\kappa},\ \bar{\ell},\ \bar{r} \in R_m \Rightarrow \bar{\kappa} + \left(\bar{\ell} + \bar{r}\right) = \bar{\kappa} + \overline{\ell + r}$$

$$= \overline{\kappa + (\ell + r)}$$

$$= \overline{(\kappa + \ell) + r}$$

$$= \overline{\kappa + \ell} + \bar{r}$$

$$= \left(\bar{\kappa} + \bar{\ell}\right) + \bar{r}$$

so, + is associative.

(ii) Let $\bar{\kappa} \in R_m$. Then

$$\bar{\kappa} + \bar{0} = \overline{\kappa + 0} = \bar{\kappa} = \overline{0 + \kappa} = \bar{0} + \bar{\kappa} = \bar{0} + \bar{\kappa}.$$

So, R_m has an identity: $\bar{0}$.

(iii) Let $\kappa \in R_m$. Then $\overline{(-\kappa)} \in R_m$ and

$$\bar{\kappa} + \overline{(-\kappa)} = \overline{\kappa + (-\kappa)} = \bar{0}$$

and

$$\overline{(-\kappa)} + \bar{\kappa} = \overline{(-\kappa) + \kappa} = \bar{0}.$$

So, each $\bar{\kappa}$ has an inverse: $\overline{(-\kappa)}$. Thus $(R_m, +)$ is a group.

(iv) Let $\bar{\kappa}, \bar{\ell} \in R_m$. Then

$$\bar{\kappa} + \bar{\ell} = \overline{\kappa + \ell} = \overline{\ell + \kappa} = \bar{\ell} + \bar{\kappa}.$$

Thus $(R_m, +)$ is a commutative group and the theorem is proved.

According to definition 9, the appropriate symbol for the inverse of $\bar{\kappa}$ is $\bar{\kappa}^{-1}$. However, since the group operation is an additive operation, it is traditional to write the inverse of $\bar{\kappa}$ as $-\bar{\kappa}$. The proof of theorem 10 then shows

$$-\bar{\kappa} = \overline{(-\kappa)}.$$

In R_3, for example,

$$-\bar{2} = \overline{(-2)} = \bar{1}$$

since $(-2) \equiv 1 \bmod 3$.

You should agree that each step leading to the statement of theorem 10, and each particular detail of its proof, is elementary. Nevertheless, the finished product—the group $(R_m, +)$—possesses a fair degree of complexity. R_m is a set that contains as elements other sets, namely, the residue classes

$\overline{0}, \overline{1}, \ldots, \overline{m-1}$. The construction of R_m required, in particular, notions of the integers, mathematical induction, well ordering, division, equivalence relation, and congruence modulo m. In addition, we defined a nontrivial binary operation on R_m in such a way that the resulting pair $(R_m, +)$ became a commutative group. Consequently, R_m—together with its binary operation—shares all the common properties contained by the rich collection of mathematical objects called Abelian groups. All things considered, $(R_m, +)$ is not a commonplace idea.

To be sure, theorem 10 is simple enough line by line. But so is Hamlet.

In passing, let's notice that we could also define a multiplicative binary operation on R_m by

$$\overline{\kappa} \cdot \overline{\ell} = \overline{\kappa\ell}.$$

Once again, we must show that the operation is well defined. But the proof is analogous to the proof of lemma 2 and the details are omitted. (The key item needed is theorem 5(ii).)

In R_3, for example, we have

$$\overline{1} \cdot \overline{2} = \overline{2}$$

and

$$\overline{2} \cdot \overline{3} = \overline{6} = \overline{0}.$$

And in R_5,

$$\overline{2} \cdot \overline{3} = \overline{6} = \overline{1}$$
$$\overline{3} \cdot \overline{3} = \overline{9} = \overline{4}.$$

Figure 29 shows complete multiplication tables for R_3 and R_5.

A straightforward check will show that the binary operation $\cdot$ for R_m is associative and commutative. Moreover, R_m contains an identity element $\overline{1}$:

$$\overline{1} \cdot \overline{\kappa} = \overline{1 \cdot \kappa} = \overline{\kappa} = \overline{\kappa \cdot 1} = \overline{\kappa} \cdot \overline{1}.$$

But $(R_m, \cdot)$ does not form a group because $\overline{0}$ has no multiplicative inverse.

R_3

•	$\bar{0}$	$\bar{1}$	$\bar{2}$
$\bar{0}$	$\bar{0}$	$\bar{0}$	$\bar{0}$
$\bar{1}$	$\bar{0}$	$\bar{1}$	$\bar{2}$
$\bar{2}$	$\bar{0}$	$\bar{2}$	$\bar{1}$

R_5

•	$\bar{0}$	$\bar{1}$	$\bar{2}$	$\bar{3}$	$\bar{4}$
$\bar{0}$	$\bar{0}$	$\bar{0}$	$\bar{0}$	$\bar{0}$	$\bar{0}$
$\bar{1}$	$\bar{0}$	$\bar{1}$	$\bar{2}$	$\bar{3}$	$\bar{4}$
$\bar{2}$	$\bar{0}$	$\bar{2}$	$\bar{4}$	$\bar{1}$	$\bar{3}$
$\bar{3}$	$\bar{0}$	$\bar{3}$	$\bar{1}$	$\bar{4}$	$\bar{2}$
$\bar{4}$	$\bar{0}$	$\bar{4}$	$\bar{3}$	$\bar{2}$	$\bar{1}$

Figure 29: Multiplication tables for R_3 and R_5

REPRISE

Early on, we looked at the prime numbers: 2, 3, 5, 7, 11, 13, ... Let's return to them for two promised results:

(1) Each natural number, larger than 1, has a unique representation as a product of primes.

(2) There exist infinitely many prime numbers.

Each of these can be traced back to 300 BCE and ancient Greece, where they appeared in Euclid's most famous work, *The Elements*. The first result—known as the *fundamental theorem of arithmetic*—appears in two parts: proposition 31 of book VII and proposition 14 of book VIII. The theorem on the infinity of primes is proposition 20 of book VIII. Euclid's proof of this stands today as a classical example of elegant mathematics. We will see it shortly.

Earlier, we anticipated the fundamental theorem by looking at the examples:

$$30 = 2 \cdot 3 \cdot 5$$

and

$$385 = 5 \cdot 7 \cdot 11.$$

Then each of the composite numbers 30 and 385 may be written as a product of primes.

Often, in such a representation, prime factors are repeated. For example,

$$360 = 2 \cdot 2 \cdot 2 \cdot 3 \cdot 3 \cdot 5$$

and

$$294 = 2 \cdot 3 \cdot 7 \cdot 7.$$

We usually write these in the canonical form

$$360 = 2^3 \cdot 3^2 \cdot 5$$

and

$$294 = 2 \cdot 3 \cdot 7^2.$$

A prime number has no positive factors except itself and 1. But if we allow the phrase "product of primes" to include a prime number standing alone, then the fundamental theorem asserts that each integer larger than 1 has such a representation. More precisely,

THEOREM 11. The fundamental theorem of arithmetic:

Let $n \in \mathbb{N}$, $n > 1$. Then there exists $m \in \mathbb{N}$ and prime numbers $p_1, p_2, \ldots,$ p_m such that

$$n = p_1 \cdot p_2 \cdot \ldots \cdot p_m.$$

Moreover, except for the order of the factors, this representation is unique.

PROOF. Let $n \in \mathbb{N}$, $n > 1$. If n is prime then the first part of the theorem is true.

Suppose n is not prime. Let S be the set of all divisors d of n that satisfy $1 < d < n$. That is,

$$S = \{d : d \in \mathbb{N}, d | n, 1 < d < n\}.$$

Since n is composite, $S \neq \varnothing$. Thus, by the well-ordering principle for the natural numbers, S contains a smallest element p_1. Then p_1 must be prime. (If p_1 is not prime then: $n = p_1 \cdot q_1$ because $p_1 | n$. Also $p_1 = rq_2$ because p_1 is composite. So, $n = rq_2q_1$. Therefore, $r \in S$ and $r < p_1$. Impossible because p_1 is the smallest element of S.)

Thus we have

$$n = p_1 q_1$$

where p_1 is prime and $1 < q_1 < n$. If q_1 is prime, then n is a product of primes. If q_1 is composite, repeat the argument (with q_1 playing the role of n). This gives:

$$q_1 = p_2 q_2$$

where p_2 is prime and $1 < q_2 < q_1$. Thus,

$$n = p_1 p_2 q_2.$$

If q_2 is prime the proof is complete. Otherwise, repeat the argument to obtain

$$n = p_1 p_2 p_3 q_3$$

with p_3 prime and $1 < q_3 < q_2$. The process must terminate in a finite number of steps because $q_1, q_2, q_3, \ldots$ are all positive integers and

$$q_1 > q_2 > q_3 > \ldots > 1.$$

When it does terminate we have

$$n = p_1 p_2 p_3 \cdots p_m$$

with $p_1, p_2, \ldots, p_m$ all prime numbers. Thus n has a representation as a product of primes.

In order to prove uniqueness, suppose two such representations exist. That is, suppose

$$n = p_1 p_2 \cdots p_m$$

and

$$n = r_1 r_2 \cdots r_k$$

where each of $p_1, p_2, \ldots, p_m$ and $r_1, r_2, \ldots, r_k$ are prime numbers. Without loss of generality we may suppose $m \leq k$.

We have

(*) $$p_1 p_2 \cdots p_m = r_1 r_2 \cdots r_k.$$

Thus, $p_1 \mid r_1 r_2, \ldots, r_k$ because $p_1 \mid p_1 p_2, \ldots, p_m$. So, by the remarks following corollary 2, $p_1 = r_i$ for some $i = 1, 2, \ldots, k$. We may then remove p_1 and r_i from both sides of equation (*). We repeat this process m times until all the p's on the left-hand side of (*) are cancelled against m of the r's on the right-hand side. The result is

(**) $$1 = s_1 s_2 \cdots s_{k-m}$$

where $s_1, s_2, \ldots, s_{k-m}$ are the r's that remain (if any) after cancellation of the p's. This gives

(***) $$1 = 1 \cdot (s_1 s_2 \cdots s_{k-m}).$$

But each $s_j > 1$ so (***) cannot hold unless none remain after cancellation of $p_1, p_2, \ldots p_m$. Hence, $k = m$ and the p's in

$$n = p_1 p_2 \cdots p_m$$

are equal one-by-one to the r's in

$$n = r_1 r_2 \cdots r_m.$$

Therefore the representations are unique and the theorem is proved.

In the literature, the conclusion of theorem 11 often appears in the form

$$n = p_1^{m_1} p_2^{m_2} \cdots p_k^{m_k}$$

where each p_j is prime, $k \in \mathbb{N}$, and each m_j is a positive integer. This notation reflects the fact the primes $p_1, p_2, \ldots, p_k$ are now distinct, equal primes having been collected and their products written in exponential form just as

$$360 = 2 \cdot 2 \cdot 2 \cdot 3 \cdot 3 \cdot 5$$

becomes

$$360 = 2^3 \cdot 3^2 \cdot 5.$$

The fundamental theorem of arithmetic tells us that primes constitute the basic elements of the natural numbers beginning with 2. Thus—in the sense of the fundamental theorem—the prime numbers together with the number 1 are more "natural" than the natural numbers. Kronecker may have been wrong when he asserted: "God made the natural numbers." Perhaps what God made is

$$1, 2, 3, 5, 7, 11, 13, 17, 19, 23, \ldots .$$

Maybe Euclid made the natural numbers—in 300 BCE, in *The Elements*. An obvious consequence of the fundamental theorem of arithmetic is:

COROLLARY 4. Each integer larger than 1 is divisible by some prime.

Corollary 4 provides the key to Euclid's elegant proof of theorem 12.

THEOREM 12. There are infinitely many primes.

PROOF. Suppose there exist only finitely many primes: $p_1, p_2, p_3, \ldots, p_n$. Let

$$m = p_1 p_2 p_3 \cdots p_n + 1.$$

By corollary 4, m is divisible by some prime. This prime cannot be p_1 because $p_1 \mid p_1 p_2, \ldots p_n$. So if $p_1 \mid m$ then $p_1 \mid (m - p_1 p_2, \ldots, p_n)$. Thus $p_1 \mid 1$, which is impossible since $p_1 > 1$. A similar argument shows that none of p_2, $p_3, \ldots, p_n$, can divide m. This contradicts our assumption that $p_1, p_2, \ldots, p_n$ are *all* the prime numbers.

Thus there are infinitely many primes.

Actually, the proof requires fewer words. Macbeth says of the weird sisters' prophecies:

> This supernatural soliciting
> cannot be ill, cannot be good.[4]

For the proof, a mathematician might simply write

$$m = p_1 p_2 \cdots p_n + 1$$

and say simply:

> This supernatural number
> cannot be prime, cannot be not.

Approached in the proper mood, Euclid's proof, like a witch in velvet, will take your breath away.

CONJECTURES

Classical number theory abounds with conjectures and open problems. Many of these are ancient enough and deep enough to bring to their conqueror mathematical fame.

Here are three open problems:

(1) *Twin prime*: If p is a prime number then $p + 1$ can be prime only if $p = 2$. (Each prime p other than 2 must be odd. So, $p + 1$ is even and cannot be prime.) But p and $p + 2$ might be prime. Examples are 3 and 5, 5 and 7, 11 and 13, 17 and 19. Such pairs of primes are called *twin primes*. *How many twin primes are there?* No one knows. Probably there are infinitely many but no proof of this exists.

(2) *Perfect numbers*: A natural number n is perfect if n equals the sum of its proper divisors. Examples are 6 and 28:

$$6 = 1 + 2 + 3$$

and

$$28 = 1 + 2 + 4 + 7 + 14.$$

The next two perfect numbers are 496 and 8,128.

How many perfect numbers are there? Do odd perfect numbers exist? No one knows.

(3) *The Goldbach conjecture*: In 1749, Christian Goldbach produced the conjecture:

Every even integer greater than 4 can be written as the sum of two odd primes.

For example,

$$6 = 3 + 3$$

$$8 = 3 + 5$$

$$10 = 3 + 7 = 5 + 5$$

$$12 = 5 + 7.$$

Computers have verified Goldbach's conjecture for all even integers between 6 and some enormous value (which keeps increasing as technology improves). But, as we have seen, verification is not mathematics. Either a proof or a counterexample must be found before Goldbach's conjecture turns true. Find either one and you can claim your own share of mathematical immortality.

Chapter 5

NUMBERS REAL AND IMAGINARY

Back somewhere in the shadowy past looms the vague figure of Pythagoras. Little is known of the details of his life, but we do know that around 550 BCE he founded his famous society in the Greek city of southern Italy called Croton. And we know that he produced Platonic-type philosophy long before Plato and that he practiced axiomatic mathematics two centuries before Euclid. Details of him may be dim but his influence is clear enough. George Simmons[1] claims that both mathematics and science begin with Pythagoras. Will Durant concurs:

> All in all, Pythagoras was the founder, so far as we know them, of both science and philosophy in Europe—an achievement sufficient for any man.[2]

I'll say. Nevertheless, Mr. Pythagoras was not without flaw. The students in his school, which bore the severe stamp of his personality and almost monastic beliefs, followed a course of study that was a curious blend of scientific precision and religious dogma. Even their mathematical notions, at least at the beginning, were influenced by their concept of the Divine. To the Pythagoreans numbers composed the essence of the universe. The principles of numbers, they believed, were the principles of everything. And by numbers they meant whole numbers and the ratios of whole numbers. Their motto "Everything Is Number" becomes in our language: "Everything Is Rational Number."

In particular, the Pythagoreans believed that all measurements could be made using only rational numbers. The gods simply would have made the world no other way. Thus an arbitrary line segment of length, say, *L* could be so measured. That is, one could find a fractional unit of measurement—call its length $1/m$—with the property that enough of these placed end to end exactly fills the line segment. If the filling requires, say, n of these units, then $L = n(1/m)$. But this means:

$$L = \frac{n}{m}$$

and the length of the given segment becomes a rational number.

MURDER FOR MEASURE

Unfortunately—at least for the Pythagoreans—it is false that $L = \frac{n}{m}$ holds for every possible length L. That is, not every length can be expressible as a rational number. Using their high-toned mathematical sophistication, the Pythagoreans discovered this fact. They might have been elated to learn that the rational numbers are not sufficiently rich to explain the universe and, consequently, motivated to construct numbers that do. But the startling discovery was at odds with their nonsensical religious notions and they were instead devastated by the new knowledge. Legend tells that a Pythagorean named Hippasus made the discovery while he and the others of the society were aboard a ship. Hippasus was particularly devastated. His pals pitched him into the sea. Out of sight, out of mind.

The mathematical details of the story fall into two parts: the second being Hippasus's result and the first, the famous Pythagorean theorem.

Consider a right triangle with sides of lengths a, b, and c, respectively, where c represents the length of the hypotenuse. Then

$$a^2 + b^2 = c^2.$$

PROOF. Construct a square with sides of length a and b as shown in figure 30. Place a point on each side of the square at distances b and a from the vertices as figure 30 shows. Now join the points pairwise in order to form four copies of the given triangle. The construction partitions the square of side $a + b$ into the four right triangles and a smaller square with sides of length c.

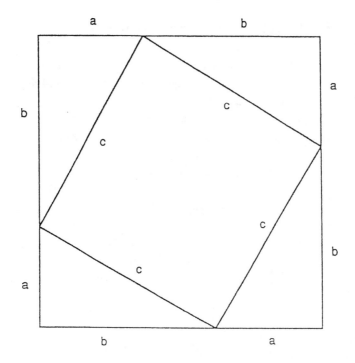

Figure 30: The Pythagorean theorem

From geometry we know that the large square has area $(a + b)^2$, the small square has area c^2, and each of the four triangles has area $\frac{1}{2}ab$. But the area of the large square equals the area of the small square plus the area of the four triangles. Hence,

$$(a+b)^2 = c^2 + 4\left(\frac{1}{2}ab\right)$$
$$(a+b)(a+b) = c^2 + 2ab$$
$$a^2 + ab + ba + b^2 = c^2 + 2ab$$
$$a^2 + 2ab + b^2 = c^2 + 2ab.$$

Therefore

$$a^2 + b^2 = c^2.$$

(The calculation $(a+b)^2 = a^2 + 2ab + b^2$ occurs often enough that it should be memorized.)

This theorem, which may have been proved by Pythagoras himself, enabled the Pythagoreans to compute the length of the diagonal of any given square, since a diagonal divides the square neatly into two right triangles. In particular, if we consider a square whose sides are of length 1 and let d denote the length of its diagonal (see figure 31), then

$$d^2 = 1^2 + 1^2$$

or

$$d^2 = 2.$$

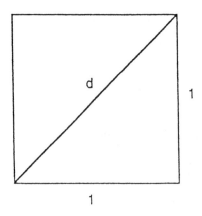

$$d^2 = 1^2 + 1^2 = 2$$

Figure 31: A diagonal whose square is 2

Thus, by his own hand, Pythagoras constructed a diagonal line segment with the property that the square of its length equals 2. There was no doubt of the existence of this segment: his mathematics told him so. And from his religion he knew that d could be expressed as some rational number.

So far so good. But then Hippasus came along—or rather went along on the murder cruise—and showed with the same Pythagorean mathematical precision that if $d^2 = 2$, then d could not be a rational number. Here's what he did:

THEOREM 1. There exists no rational number d with

$$d^2 = 2.$$

PROOF. Suppose d is rational. That is, suppose

$$d = \frac{n}{m}$$

with $n, m \in \mathbb{Z}$ and $m \neq 0$. We know that common factors may be cancelled from the numerator and denominator of a rational number. So, we suppose this has already been done so that n and m contain no common factors. Hence, our assumption takes the form:

$$d = \frac{n}{m}$$

with $n, m \in \mathbb{Z}$ and $gcd\ (n, m) = 1$.

Then

$$d^2 = 2 \Rightarrow \left(\frac{n}{m}\right)^2 = 2$$
$$\Rightarrow n^2 = 2m^2$$
$$\Rightarrow 2 \mid n^2.$$

Thus n^2 is even. Then n must be even because the square of an odd integer is odd. (If r is odd then $r = 2q + 1$ for some $q \in \mathbb{Z}$. Then $r^2 = (2q + 1)^2 = 4q^2 + 4q + 1 = 2(2q^2 + 2q) + 1 = 2Q + 1$ is also odd.)

Hence, $2 \mid n$ so $n = 2s$ for some $s \in \mathbb{Z}$. Then

$$n^2 = 2m^2 \Rightarrow (2s)^2 = 2m^2$$
$$\Rightarrow 4s^2 = 2m^2$$
$$\Rightarrow 2s^2 = m^2$$
$$\Rightarrow 2 \mid m^2.$$

So, m^2 is even and, as before, m must be even. Hence, $2 \mid m$. Thus we have $2 \mid n$ and $2 \mid m$, which contradicts our assumption that $gcd\ (n, m) = 1$. Therefore, d is not a rational number.

Pythagoras and his students then faced what must have seemed an almost equally balanced dilemma: here stands a clearly defined number d. Our religion tells us d is rational but our mathematics tells us it is not. What to do?

The Pythagoreans chose simply to consider the dilemma as bad news. They murdered the bearer of the news and, for a time, cooperatively hid away his message, thus producing an ancient example of the modern political concept of a cover-up. Later, however, mathematics prevailed over mystagogy and they came to terms with Hippasus's theorem. Bertrand Russell tells us that the Pythagoreans then went so far as to develop methods for successive approximation of the elusive number whose square is 2.[3]

The significance of theorem 1 lies not only in its importance—showing as it does the need for numbers beyond the rationals—but also in its elegance. Stripped of explanatory comments, the entire proof occupies only five lines:

Suppose $\left(\dfrac{n}{m}\right)^2 = 2$, where $n, m \in \mathbb{Z}$ and $gcd\,(n, m) = 1$.

Then $n^2 = 2m^2$, so that $2 \mid n^2$ and thus $2 \mid n$.

So, $n = 2r$ for some $r \in \mathbb{Z}$.

This gives $4r^2 = 2m^2$ so that $2 \mid n^2$ and thus $2 \mid m$.

But $2 \mid n$ and $2 \mid m$ contradicts $gcd\,(n, m) = 1$.

In *A Mathematician's Apology*, G. H. Hardy gives this result and Euclid's proof of the infinitude of the prime numbers as examples of theorems of "the highest class." Hardy said of them:

Each is as fresh and significant as when it was discovered—two thousand years have not written a wrinkle on either of them.[4]

Yes.

THE REAL LINE

The number d of theorem 1 has the property that $d^2 = 2$ and, since d is the length of a diagonal, $d > 0$. Thus d is a positive number whose square is 2. Suc, a number—if it exists—is called the *square root of 2* and is denoted by $d = \sqrt{2}$. In this language the statement of theorem 1 becomes

$$\sqrt{2} \text{ is not a rational number.}$$

Theorem 1 tells us that if we want to solve the simple equation $x^2 = 2$, that is, to determine $\sqrt{2}$, we must look somewhere beyond the rational numbers.

Nowhere in $\mathbb{Q}$ will we find $\sqrt{2}$.

Let's look at the situation geometrically. Figure 32 shows a line on which certain points have been marked. The (arbitrarily chosen) point marked 0 is called the *origin*. A convenient unit of length was selected and the integer 1 was marked at this distance to the right of the origin. Then, using the same unit repeatedly, the integers 2, 3 , 4, . . . were located farther to the right. Analogously, the negative integers −1, −2, −3, . . . are positioned to the left of the origin.

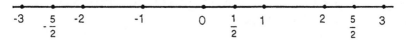

Figure 32: The rational number line

Figure 32 also shows certain rational numbers such as $\dfrac{1}{2}$, $\dfrac{5}{2}$, and $-\dfrac{5}{2}$. These rationals, of course, are located according to their distance to the right or left of the origin. Thus any rational number appears somewhere on this line, positive rationals to the right of the origin and negative rationals to the left.

$$r_1 = \frac{p+q}{2}$$

$$r_2 = \frac{p+r_1}{2}$$

$$r_3 = \frac{p+r_2}{2}$$

. . .

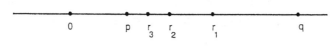

Figure 33: Density of the rational numbers

Moreover, the rational numbers lie *densely* on the line. By this we mean the following:

Let $p, q \in \mathbb{Q}$ with $p < q$. (See figure 33.) Let

$$r_1 = \frac{p+q}{2}.$$

Then $r_1 \in \mathbb{Q}$ and, being the average of p and q, falls between them: $p < r_1 < q$. ($p < q \Rightarrow p + q < 2q \Rightarrow \dfrac{p+q}{2} < q$. Thus $r_1 < q$. Similarly, $p < r_1$.) Now let

$$r_2 = \frac{p + r_1}{2}$$

so that r_2 falls between p and r_1. Continuing in this manner we obtain an infinite sequence of rational numbers r_1, r_2, r_3, r_4, ... with the property

$$q > r_1 > r_2 > r_3 > \ldots > p.$$

Thus, between any two rational numbers there lives an infinity of other rational numbers.

But, in spite of the density of the rational numbers, they still do not completely fill the line of figure 33. Holes remain. In particular, theorem 1 says that no rational number occupies the point whose distance to the right of the origin is precisely d, when $d^2 = 2$. That is, there exists no rational number whose distance from the origin equals $\sqrt{2}$.

Moreover, many other holes remain. It can be shown, in a manner analogous to the proof of theorem 1, that there exists no rational number r such that $r^2 = 3$; that is, $\sqrt{3}$ is not rational. Similarly, $\sqrt{5}$, $\sqrt{7}$, $\sqrt{11}$, ... are not rational. From this, it follows that many combinations such as $\dfrac{\sqrt{2}}{2}$, $\dfrac{1}{2} + \sqrt{3}$, and so on fail to be rational numbers. So, in terms of only rational numbers, the line of figure 33 is as sliced with holes as Caesar was with wounds.

Technically, it makes no sense for us—at this point—even to speak of numbers like $\sqrt{2}$. At this time, we have before us only the rational numbers $\mathbb{Q}$, and we know that $\sqrt{2}$ cannot be rational. Thus—if we were proceeding with complete rigor—we would now be compelled to enlarge our number system in such a manner as to produce a new number d such that $d^2 = 2$. The precise process of enlargement constitutes the next stage in the Peano program of number construction. When this has been done, all the holes in the line of figure 30 are filled, either with rational numbers such as $\frac{5}{2}$ or else with new numbers such as $\sqrt{2}$. The new numbers are called *real numbers* and the line of figure 32 becomes the *real number line* of figure 34.

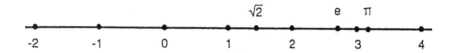

Figure 34: The real number line

The actual construction of the real numbers from the rational numbers requires a higher level of sophistication than we want to pursue in this book and we will not bother with details. (See Landau's book for a full presentation.)[5] But the basic notion consists of the adjunction to $\mathbb{Q}$ of new numbers that can serve as the limits of certain sequences of rational numbers. It is necessary—before the process begins—to define precisely the concept of *sequence* and of *limit*. We will, however, content ourselves with an intuitive discussion of a single illustration.

We know by theorem 1, that $d^2 = 2$ holds for no $d \in \mathbb{Q}$. The line of figure 32 has a hole at this position. Part of the construction process then must consist of determining exactly where such a number should be located on the line of figure 32, the real number line. Since $1^2 = 1 < 2$ and $2^2 = 4 > 2$ we see that d must fall between 1 and 2. We come upon it by successive approximations.

Consider $c_1 = 1.4$ and $d_1 = 1.5$. Notice that $c_1, d_1 \in \mathbb{Q}$ because $1.4 = \frac{14}{10}$ and $1.5 = \frac{15}{10}$. Then

$$c_1^2 = (1.4)^2 = 1.96 < 2$$

and

$$d_1^2 = (1.5)^2 = 2.25 > 2.$$

Thus, d must fall between c_1 and d_1 so that $c_1 < d < d_1$ or $1.4 < d < 1.5$.

Now consider $c_2 = 1.41$ and $d_2 = 1.42$. Again $c_1, d_2 \in \mathbb{Q}$ because

$$c_1 = \frac{141}{100} \text{ and } d_2 = \frac{142}{100}.$$

Then

$$c_1^2 = (1.41)^2 = 1.9881 < 2$$

and

$$d_1^2 = (1.42)^2 = 2.0164 > 2.$$

This tells us

$$1.41 < d < 1.42.$$

Continuing in this manner (after appropriate experimentation) we obtain an infinite sequence of rational numbers

(a) $1, 1.4, 1.41, 1.414, 1.4142, \ldots$

Each of these has, as its squares, a rational number less than 2. But the sequence of squares:

$$1, 1.96, 1.9881, 1.999396, 1.99996164, \ldots$$

becomes arbitrarily close to 2 as we go out farther and farther. Thus we *define* the square root of 2 to be the *infinite decimal* produced as the limit of sequence (a). It turns out that

$$d = \sqrt{2} = 1.41421356 \ldots .$$

The process will be meaningful only *after* the concepts of sequence and limit are precisely defined. Moreover, the program must be carried out in sufficient generality so that it yields not only $d = \sqrt{2}$ but also produces expressions for distances such as r and ℓ where

$$r^2 = 3 \text{ and } \ell^2 = 5$$

since they also make positions where holes exist in the rational number line of figure 32. Accomplishing all this is no small task and frankly lies beyond the scope of this book. But the preceding discussion illustrates the gist of an acceptable method for doing the job. When all is done, we have before us a new set of numbers. We call these the *real numbers*. The totality of these numbers—the rationals and the irrationals—completely fills the holes in the rational line of figure 32. The resulting line now has a precisely defined real number corresponding to each of its points. Figure 34 shows this new *real number line*.

REAL NUMBER OPERATIONS

The set of real numbers is traditionally denoted by $\mathbb{R}$. Often, the real number line of figure 34 is called simply the real line. By means of its construction, $\mathbb{R}$ contains each rational number as an element. Thus, $1 \in \mathbb{R}$ and $\frac{5}{2} \in \mathbb{R}$. But, as we have seen, $\mathbb{R}$ contains many elements—such as $\sqrt{2}$ and $\sqrt{3}$ —which do not belong to $\mathbb{Q}$. Thus we have the growing chain of sets:

$$\mathbb{N} \subset \mathbb{Z} \subset \mathbb{Q} \subset \mathbb{R}$$

where each of the inclusions are proper inclusions.

The method of construction of $\mathbb{R}$ allows the new set to inherit from its proper subset $\mathbb{Q}$ the properties of addition, multiplication, and order. Once the appropriate definitions are given for $\mathbb{R}$ a theorem formally identical to theorem 16 of chapter 3 can be proved. That is, each statement of theorem 16 may be replaced by an analogous statement for the larger set $\mathbb{R}$. All we need do is replace the symbol $\mathbb{Q}$, whenever it appears, by $\mathbb{R}$.

THEOREM 2. The statement and conclusions of theorem 16 of chapter 3 hold for the real numbers $\mathbb{R}$.

The formal equivalence between theorem 16 of chapter 3 and theorem 2 above does not imply that $\mathbb{R}$ and $\mathbb{Q}$ are themselves identical. We know, for instance, that $\mathbb{Q}$ is a proper subset of $\mathbb{R}$. It turns out that in $\mathbb{R}$ we can always solve the equation $x^2 = c$ wherever $c > 0$. The solutions are $x = \pm\sqrt{c}$. (Theorem 1 asserts the impossibility of this for $\mathbb{Q}$ even in the simple case $c = 2$.) And the real numbers also possess another distinguishing property. The real numbers are *complete*.

A true discussion of the notion of *completeness* would take us far afield. But, roughly, it goes like this:

> A set of numbers is called *complete* if each sequence of numbers belonging to the set that is trying to converge actually converges.

> For example, the sequence of members of $\mathbb{Q}$:

$$1, 1.4, 1.41, 1.414, 1.4142, \ldots$$

> seems to tend to a number d with the property $d^2 = 2$. But no such number lives in $\mathbb{Q}$. Thus, $\mathbb{Q}$ *is not complete.*
> The same sequence, however, does converge in $\mathbb{R}$ because $d = \sqrt{2}$ and $\sqrt{2} \in \mathbb{R}$. $\mathbb{R}$ is complete precisely because each such sequence converges to an element of $\mathbb{R}$.

In a sense that we cannot pursue in this book, the properties described in theorem 2 together with the concept of completeness characterize the real numbers. No other set in the mathematical world satisfies all these conditions. In this sense, the real numbers are unique.

Just as theorem 16 of chapter 3 provides techniques for the manipulation of rational numbers, theorem 2 allows us to make similar calculations with members of $\mathbb{R}$. But we must be careful when we deal with square roots. We cannot, for example, always solve the equation $x^2 = y$ in $\mathbb{R}$. This follows from the next theorem:

THEOREM 3. If $x \in \mathbb{R}$, $x \neq 0$, then $x^2 > 0$.

PROOF. If $x > 0$ then $x \cdot x > x \cdot 0$ or $x^2 > 0$ by theorem 2, part O(iv). ($x \cdot 0 = 0$ holds exactly as it does for the rational numbers. The proof is identical.) If $x < 0$ then $x = -a$ when $a > 0$. Thus $x^2 = (-a)(-a) = a^2 > 0$. ($(-a)(-b) = (a)(b)$ holds for real numbers exactly as it does for rational numbers.)

Since $0^2 = 0 \cdot 0 = 0$, theorem 3 tells us that, if $x \in \mathbb{R}$ then $x^2 \geq 0$. Thus *there exists no real number* x *with* x^2 = y *if* y < 0. In particular, there exists no real number x such that $x^2 = -1$.

Consequently, the symbol $\sqrt{a}$ makes sense in $\mathbb{R}$ only when $a \geq 0$. Moreover, $\sqrt{a}$ itself is, by definition, always positive or zero. In summary,

$\sqrt{0} = 0$ and if $a > 0$ then $\sqrt{a}$ is the unique positive real number with $(\sqrt{a})^2 = a$.

In particular $\sqrt{4} = 2$ because $2 > 0$ and $2^2 = 4$. Notice, however, that if $x^2 = 4$ then $x = \pm\sqrt{4} = \pm 2$:

$$x^2 = 4 \Rightarrow x^2 - 4 = 0$$
$$\Rightarrow (x - 2)(x + 2) = 0$$
$$\Rightarrow x - 2 = 0 \text{ or } x + 2 = 0$$
$$\Rightarrow x = 2 \text{ or } x = -2.$$

The next-to-last implication follows from the fact that

$$u, w \in \mathbb{R} \text{ and } u \cdot w = 0 \Rightarrow u = 0 \text{ or } w = 0$$

which comes easily from theorem 2, part M(v).

The preceding calculation extends easily to the following:

THEOREM 4. Let $y \in \mathbb{R}$ and $y > 0$. If $w^2 = y$ then $w = \pm\sqrt{y}$.

PROOF.
$$w^2 = y \Rightarrow w^2 - y = 0$$
$$\Rightarrow \left(w - \sqrt{y}\right)\left(w + \sqrt{y}\right) = 0$$
$$\Rightarrow w - \sqrt{y} = 0 \text{ or } w + \sqrt{y} = 0$$
$$\Rightarrow w = \sqrt{y} \text{ or } w = -\sqrt{y}.$$

So the equation $w^2 = y$ has the two solutions $w = \pm\sqrt{y}$ if $y > 0$. (The "factorization" following the second implication is a special case of $w^2 - u^2 = (w - u)(w + u)$.)

If we extend the calculation once more we obtain the famous:

THEOREM 5. The quadratic formula: Let $a, b, c \in \mathbb{R}$, $a > 0$, and $b^2 - 4ac \geq 0$. If $ax^2 + bx + c = 0$ then

$$x = \frac{-b \pm \sqrt{b^2 - 4ac}}{2a}.$$

PROOF.

$$ax^2 + bx + c = 0 \Rightarrow ax^2 + bx = -c$$

$$\Rightarrow a\left(x^2 + \frac{b}{a}x\right) = -c$$

$$\Rightarrow x^2 + \frac{b}{a}x = -\frac{c}{a}$$

$$\Rightarrow x^2 + \frac{b}{a}x + \frac{b^2}{4a^2} = -\frac{c}{a} + \frac{b^2}{4a^2}$$

$$\Rightarrow \left(x + \frac{b}{2a}\right)\left(x + \frac{b}{2a}\right) = \frac{b^2}{4a^2} - \frac{c}{a}$$

$$\Rightarrow \left(x + \frac{b}{2a}\right)^2 = \frac{b^2}{4a^2} - \frac{c}{a} \cdot \frac{4a}{4a}$$

$$\Rightarrow \left(x + \frac{b}{2a}\right)^2 = \frac{b^2 - 4ac}{4a^2}$$

$$\Rightarrow x + \frac{b}{2a} = \pm\sqrt{\frac{b^2 - 4ac}{4a^2}}$$

$$\Rightarrow x + \frac{b}{2a} = \frac{\pm\sqrt{b^2 - 4ac}}{2a}$$

$$\Rightarrow x = -\frac{b}{2a} \pm \frac{\sqrt{b^2 - 4ac}}{2a}$$

$$\Rightarrow x = \frac{-b \pm \sqrt{b^2 - 4ac}}{2a}.$$

The validity of the fourth implication from the bottom follows from theorem 4 with

$$w = x + \frac{b}{2a} \text{ and } y = \frac{b^2 - 4ac}{4a^2}.$$

The third implication from the bottom uses the fact that $\sqrt{\dfrac{u}{v}} = \dfrac{\sqrt{u}}{\sqrt{v}}$ if $u \geq 0$ and $v > 0$. (The proof of this is left as an exercise.) Mathematicians refer to the trick of adding $\dfrac{b^2}{4a^2}$ to each side in the fourth line of the proof as "completing the square."

An expression of the form $p(x) = ax^2 + bx + c$ is known as a *polynomial of second degree*. A value of x for which $p(x) = 0$ is called a *root* (or a *zero*) of the polynomial. The quadratic formula gives us—under certain conditions—specific values for the roots. (The equation $ax^2 + bx + c = 0$ is called a *quadratic equation*.)

Consider, for example, $p(x) = 2x^2 - 9x - 5$. Here, $a = 2$, $b = -9$, $c = -5$. So,

$$p(x) = 0 \Rightarrow 2x^2 - 9x - 5 = 0$$

$$\Rightarrow x = \frac{-(-9) \pm \sqrt{(-9)^2 - 4(2)(-5)}}{2(2)}$$

$$\Rightarrow x = \frac{9 \pm \sqrt{81 + 40}}{4}$$

$$\Rightarrow x = \frac{9 \pm \sqrt{121}}{4}$$

$$\Rightarrow x = \frac{9 \pm 11}{4}$$

$$\Rightarrow x = \frac{9 + 11}{4} \text{ or } x = \frac{9 - 11}{4}$$

$$\Rightarrow x = 5 \text{ or } x = -\frac{1}{2}.$$

The condition $a > 0$ in the quadratic formula can be replaced by the weaker requirement $a \neq 0$. (If $a < 0$, just multiply $ax^2 + bx + c = 0$ on both sides by (-1)). If $a = 0$, the equation becomes $bx + c = 0$, which has the solution $x = -\dfrac{c}{b}$ if $b \neq 0$.) But $b^2 - 4ac \geq 0$ is critical. If $b^2 - 4ac < 0$ then $\sqrt{b^2 - 4ac}$ does not exist in the real numbers. (Theorem 3.) Thus if $b^2 + 4ac < 0$ then the quadratic formula fails in $\mathbb{R}$. The quantity $b^2 - 4ac$ is known as the *discriminant* of the quadratic equation $ax^2 + bx + c = 0$. (This is the definition of discriminant.)

AN UNCOUNTABLE INFINITY

It follows from the method of construction of $\mathbb{R}$ that each real number can be represented by an infinite decimal. Conversely, any infinite decimal represents some real number. For example,

$$2 = 2.000 \ldots$$

$$\frac{1}{2} = .5 = .5000 \ldots$$

$$\frac{1}{3} = .333 \ldots$$

$$\sqrt{2} = 1.41421356 \ldots.$$

Such representations are obvious for the integers. For the other rationals, the expansions can be obtained by the ordinary process of long division. The expansions for irrational numbers such as $\sqrt{2}$ may be found by methods of successive approximation. It is a fact (which we cannot prove at this point) that the decimal expansion for each rational number ends with a string of zeros (we might say the decimal "terminates") or else closes with an infinitely repeating pattern. For example,

$$\frac{1}{2} = .5000 \ldots$$

$$\frac{1}{3} = .333 \ldots$$

$$\frac{13}{7} = 1.857142857142857142 \ldots.$$

Moreover, it can be shown that any such repeating decimal represents a rational number. Thus the decimal representation for each irrational number fails to have an infinitely repeating pattern.

There is a subtlety here that we can only handle descriptively: *the decimal expansions for the real numbers are not unique. A string of zeros essentially may be replaced by a string of nines.*

For example,

$$1 = 1.0000 \ldots$$

and

$$1 = .9999 \ldots.$$

These two representations differ in that any termination of the first yields a finite decimal that exactly equals 1 while a similar termination of the second produces a real number less than 1. For example,

$$1 = 1.00000$$

but

$$1 > .99999.$$

Nevertheless, the second expression, $1 = .999 \ldots$, produces an exact equality as long as we interpret the three dots to mean that the nines go on forever. The real number $.999 \ldots$ is precisely equal to 1, and no approximation is involved.

A rigorous proof of this fact lies beyond the scope of this book since it involves a sophisticated application of the completeness property of the real numbers. However, we can convince ourselves of its validity if we assume (correctly in this case) that we can manipulate the infinite decimal as we would an ordinary decimal:

Let

$$x = .9999 \ldots.$$

Then

$$10x = 9.999 \ldots.$$

Hence,

$$10x - x = 9.999 \ldots - .9999 \ldots$$
$$= 9.$$

So,

$$9x = 9 \text{ or } x = 1.$$

Thus,

$$1 = .999 \ldots.$$

As we change the decimal representation of a real number from one that ends in zeros to another that ends in nines, we may have to modify another digit. For example,

$$6.243000\ldots = 6.242999\ldots.$$

But, if we decide in advance to choose one representation or another, then we may assert that each real number has a unique decimal representation and, conversely, each such decimal expansion represents a unique real number. In particular, if x is a real number and $0 < x < 1$ then x has a unique representation of the form

$$x = .b_1b_2b_3\ldots$$

where each b_k is one of $0, 1, 2, \ldots, 9$.

With these preliminaries behind, we are now able to fulfill the promise made in chapter 2 and show that there exists no one-to-one correspondence between the infinite sets $\mathbb{N}$ and $\mathbb{R}$.

THEOREM 6. There exists no one-to-one correspondence between the set of natural numbers and the set of real numbers.

PROOF. We shall show that no such correspondence exists because the set $\mathbb{R}$ is too large. And we will accomplish this by showing that the proper subset $S \subset \mathbb{R}$ defined by

$$S = \{x : x \in \mathbb{R},\ 0 < x < 1\}$$

is by itself too large to be placed in a one-to-one correspondence with $\mathbb{N}$. The method used is the famous diagonal argument of Georg Cantor (1845–1918).

Suppose there exists a one-to-one correspondence between S and $\mathbb{N}$. This means that the elements of S can be written in a list

$$x_1, x_2, x_3, \ldots$$

corresponding to the list of natural numbers $1, 2, 3, \ldots$.

Let us now write this list of elements of S in a vertical column and associate with each x_k its unique decimal expansion. (In order not to run out of symbols, we must use a system of double indices.) This gives

$$x_1 = .a_{11}a_{12}a_{13}\ldots$$

$$x_2 = .a_{21}a_{22}a_{23} \cdots$$

$$x_3 = .a_{31}a_{32}a_{33} \cdots$$

$$\cdots,$$

where each digit a_{nm} is either 0, 1, 2, . . . , 9. Now let y be the real number defined by

$$y = .b_1b_2b_3, \cdots$$

where each b_k is either 0, 1, 2, . . . , 9 but $b_1 \neq a_{11}$, $b_2 \neq a_{22}$, $b_3 \neq a_{33}$. Then

$$y \in \mathbb{R} \text{ and } 0 < y < 1.$$

But $y \neq x_1$ because its first digit differs from the first digit in the decimal expansion of x_1. Also, $y \neq x_2$ because its second digit differs from the second digit of x_2. Similarly, $y \neq x_3$, $y \neq x_4$, and so on. Thus y is an element of S but it appears nowhere in the list of elements of S.

Consequently, no one-to-one correspondence between $\mathbb{N}$ and S can exist. Any attempt will always omit at least one element.

> (Thus the one-to-one correspondence listed above could have only been a correspondence between $\mathbb{N}$ and a proper subset of S.) But S is a proper subset of $\mathbb{R}$. Hence $\mathbb{R}$ contains at least as many elements as S. Thus no one-to-one correspondence between $\mathbb{N}$ and $\mathbb{R}$ can exist.

This proves the theorem.

Any set that can be placed in a one-to-one correspondence with the set of natural numbers is called a *countable set*. Theorem 5 can be stated as:

The set of real numbers is not countable.

The proof of theorem 6 actually shows that $S = \{x : x \in \mathbb{R}, 0 < x < 1\}$ is not countable. S is an example of a set of real numbers called an *interval* and in terms of the real line of figure 32 represents a line segment of length 1. Since we know from chapter 2 that any two segments contain the same number of points, it follows:

Each interval $I = \{x : x \in \mathbb{R}, a < x < b\}$ is uncountable, where "uncountable" simply means "not countable."

What about $\mathbb{Q}$? The rational numbers live between the natural numbers and the real numbers: $N \subset \mathbb{Q} \subset \mathbb{R}$. N is countable and $\mathbb{R}$ is uncountable. How large is $\mathbb{Q}$?

It turns out that $\mathbb{Q}$ is countable and thus contains the same numbers of elements as does N. One method of proof of this fact is to show that the elements of $\mathbb{Q}$ can be written in a doubly infinite array, with the first row containing all the rational numbers whose denominators (in reduced form) are 1, the second those whose denominators are 2, and so forth. The second step of the proof consists of showing that this doubly infinite array can be written as a simple list of the form $x_1, x_2, x_3, \ldots$. Details are left as an exercise.

π AND ε

Figure 32 shows as points on the real line the numbers π and e. These numbers play such significant roles in mathematics and applications of mathematics that they deserve a special look. π is the best known of the two and we consider it first.

By definition, a circle is the set of points that are at a fixed distance r from a given point P. The given point is called the *center* of the circle and the fixed distance is called the *radius* of the circle. Thus the radius r of a circle is some real number. The number $d = 2r$ is known as the *diameter* of the circle. (See figure 35.)

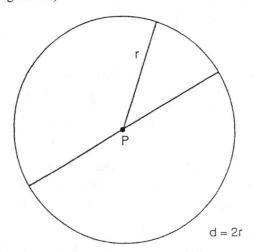

Figure 35: A circle with radius r and distance d

Now consider two arbitrary circles with radii r_1 and r_2 and diameters d_1 = $2r_1$ and $d_2 = 2r_2$, respectively. We assume that the concept of *circumference* of a circle is known to us (at least intuitively). Thus we assume familiarity with the notion of length as applied to circles: the circumference of a circle being just the ordinary distance covered by a particle moving once around it. Let C_1 and C_2 denote, respectively, the circumferences of our two circles. (See figure 36.)

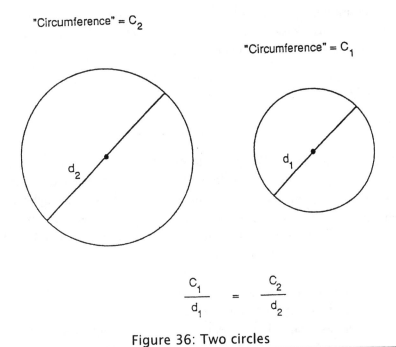

$$\frac{C_1}{d_1} = \frac{C_2}{d_2}$$

Figure 36: Two circles

The circles of figure 36 are completely arbitrary—think of one as huge as a circular lake and the other as small as a dime. But no matter what may be the variance in their sizes, the following theorem (which we shall not prove) expresses a remarkable common property—namely, that the circumferences and diameters are proportional.

THEOREM. 7 Let d_1, C_1 and d_2, C_2 denote, respectively, the diameters and circumferences of two given circles. Then

$$\frac{C_1}{d_1} = \frac{C_2}{d_2}.$$

Thus the ratio of the circumference C of an arbitrary circle to its diameter d does not depend on either C or d but is always a constant value. Mathematicians the world over denote this constant value by π, the sixteenth letter of the Greek alphabet. Hence,

$$\frac{C}{d} = \pi.$$

So, by definition, *the number π is the ratio of the circumference of an arbitrary circle to its diameter*, this ratio being constant by theorem 7. The definition often appears in the equivalent forms:

$$C = \pi d$$

or

$$C = 2\pi r.$$

Long before circles became well-defined mathematical objects humans studied their real-world approximations. From these early approximations (perhaps figures drawn in sand by a stake on the end of a tied rope) rough measurements of circumferences and diameters were made. The practical geometers of antiquity observed the apparent invariance of $\frac{C}{d}$ and estimated its value. One of the earliest estimates of the value of π was 3.

> And he made a molten seas, ten cubits from the one brim
> to the other: it was round all about, and its height was five
> cubits: and a line of thirty cubits did compass it round
> about.[6]

That was the work of Hiram, king of Tyre in 969 BCE. Hiram cut a deal with Solomon and provided him with cedar and pine for the sacred temple. Without Hiram, the job could not have been done. Hiram wrought great work. But he found an erroneous value for π:

$$\frac{C}{d} = \frac{30}{10} = 3.$$

We know now that π is an irrational number and, consequently, cannot be represented by a terminating decimal or one with an infinitely repeating pattern. The decimal expansion begins like this

$$\pi = 3.14159265358979323846\ldots.$$

The expansion is known to many thousands of places. (As I write, I have before me a computer printout giving the first ten thousand decimal places.) But the mathematics tells us that there are infinitely many digits and that there is no repeating pattern. None of us then will ever know them all.

In some ways the number e plays a complementary mathematical role to π. The two often stand together in mathematical expressions, particularly in the formulas and equations of higher analysis. For example, e and π both appear in the equation that describes the well-known normal distribution—the bell-shaped curve that dominates much of probability and statistics. However, e fails to have an elementary geometric interpretation, as does π through its connection with circumferences of circles. Unlike π, a proper definition of e requires some type of appeal to the completeness property of the real numbers. A standard method is to define e (just as $\sqrt{2}$ was defined) as the limit of an appropriate sequence of rational numbers.

Consider the sequence:

(b) $$\left(1+1\right)^1,\ \left(1+\frac{1}{2}\right)^2,\ \left(1+\frac{1}{3}\right)^3,\ \left(1+\frac{1}{4}\right)^4,\ \ldots,\ \left(1+\frac{1}{n}\right)^n\ldots.$$

Here the expression $x_n = \left(1+\dfrac{1}{n}\right)^n$ denotes the general term of the sequence.

Thus, the thousandth term is

$$x_{1,000} = \left(1+\frac{1}{1,000}\right)^{1,000}$$

and the millionth term is

$$x_{1,000,000} = \left(1+\frac{1}{1,000,000}\right)^{1,000,000}.$$

My pocket calculator gives the following values for these six terms of sequence (b):

$$x_1 = \left(1+1\right)^1 = 2$$

$$x_2 = \left(1+\frac{1}{2}\right)^2 = 2.25$$

$$x_3 = \left(1 + \frac{1}{3}\right)^3 = 2.3704$$

$$x_4 = \left(1 + \frac{1}{4}\right)^4 = 2.4414$$

. . .

$$x_{1,000} = \left(1 + \frac{1}{1,000}\right)^{1,000} = 2.71692$$

. . .

$$x_{1,000,000} = \left(1 + \frac{1}{1,000,000}\right)^{1,000,000} = 2.7182805.$$

The billionth term of the sequence is

$$x_{1,000,000,000} = \left(1 + \frac{1}{1,000,000,000}\right)^{1,000,000,000} = 2.718281827.$$

Thus the terms of the sequence appear to be tending to a real number whose value is approximately 2.7182. The sequence does in fact converge in a rigorous, mathematical sense and e is defined to be the value to which it converges. Mathematicians abbreviate all this by writing

$$e = \lim_{n \to \infty} \left(1 + \frac{1}{n}\right)^n$$

which is read: "e equals the limit as n approaches infinity of $\left(1 + \frac{1}{n}\right)^n$."

The number e has been known to be irrational since the days of Leonhard Euler (1707–1783) and, consequently, e has no terminating or repeating decimal representation. However, the decimal expansion for e has been computed to high accuracy. The first fifteen digits after the decimal are:

$$e = 2.718281828459045. \ldots$$

We shall learn shortly that much of mathematical analysis involves the concept of a *function of a real variable*. Often associated with a given function there is another function called its *derivative*. The concept of derivative lives at the deep heart's core of calculus and all mathematics that grows from calculus. When Fermat searched for an understanding of the notion of tangent line, he was actually struggling with the vague idea of derivative. When Newton wanted to express the concepts of velocity and acceleration in mathematical form, he created the object now called the derivative of a function.

Part of the mathematical significance of the number e stems from its role with respect to derivatives. The derivative of a function is another function. (For example, the derivative of $f(x) = x^2$ is $g(x) = 2x$.) It is natural to ask: "When is a function equal to its derivative?"

The surprising answer involves the number e. It can be shown that the only functions in the entire mathematical world that are identical to their derivatives are multiples of the *exponential function e^x*.

THE COMPLEX NUMBERS

In the beginning we could only count, and then only by pointing and speaking: "1, 2, 3, ..." Now we have before us numbers too numerous to count—so many they completely fill a line.

The real numbers provide us with an exceedingly rich mathematical structure: two binary operations, an ordering, and a completeness property that ensures each sequence that attempts to converge actually converges. The completeness property allows the construction of irrational numbers such as $\sqrt{2}$ and e. And the abundance of nice relationships among the binary operations and the ordering enables us to do arithmetic and algebraic calculations that were not operations and the ordering enables us to do arithmetic and algebraic calculations that were not possible within the integers or even the rational numbers. We can do lots of mathematics with the real numbers. But not quite enough. The program that began with the Peano axioms and that led us to the reals does not end just yet. One more step remains.

We know from theorem 3 that if x is any real number, then $x^2 \geq 0$. Thus we cannot—within the reals—solve the equation

$$x^2 = -1$$

because there exists no real number whose square is negative. Moreover, the quadratic formula provides solutions to the equation $ax^2 + bx + c = 0$ only when $b^2 - 4ac \geq 0$. The quadratic formula fails if $b^2 - 4ac < 0$, and we can see from the proof that this failure comes precisely from the fact that $x^2 \geq 0$ for each real number x. Therefore, if we want to solve quadratic equations whose discriminant is negative (and solve equations of higher degree), we must further enlarge our number system. In particular, we must produce a legitimate mathematical object whose square is -1.

When the enlargement is properly done, and appropriate definitions of addition and multiplication are set down, we obtain the complex numbers.

It would take us too far afield to give a careful illustration to the development of these numbers. Moreover, the truly interesting properties of the complex numbers and their significant applications become intelligible only in the presence of mathematical techniques and machinery beyond the reach of this book. We will content ourselves, therefore, with only a cursory look at these numbers and at the process that produces them.

One way to introduce the complex numbers is to begin with the set of all pairs of real numbers:

$$\mathbb{C} = \{(x, y) : x, y \in \mathbb{R}\}.$$

Then we define on $\mathbb{C}$ operations of addition and multiplication as follows:

(c) $$(x_1, y_1) + (x_2, y_2) = (x_1 + x_2, y_1 + y_2)$$

and

(d) $$(x_1, y_1)(x_2, y_2) = (x_1 x_2 - y_1 y_2, x_1 y_2 + x_2 y_1).$$

(Notice again the dual use of the symbols for addition and multiplication.)

The result of either of these operations is another pair of real numbers and, consequently, is an element of $\mathbb{C}$. Thus, addition and multiplication are binary operations on $\mathbb{C}$. One can prove directly from the definitions that addition and multiplication are each commutative and associative and that multiplication distributes over addition.

The next step consists of identifying each element of $\mathbb{C}$ of the form $(x, 0)$ with the real number x. Thus we write

$$(x, 0) = x.$$

This identification turns out to be *consistent* with the operations of addition and multiplication on both $\mathbb{C}$ and $\mathbb{R}$ in the sense that we obtain identical results. We add or multiply $(x, 0)$ and $(y, 0)$ as either elements of $\mathbb{C}$ or elements of $\mathbb{R}$. For example:

$$(x,\ 0)\cdot(y,\ 0) = (xy - 0\cdot 0,\ 0\cdot y + x\cdot 0)$$
$$= (xy,\ 0)$$

when multiplication is performed in $\mathbb{C}$. But the identification gives

$$(xy,\ 0) = xy$$

which is just multiplication in $\mathbb{R}$. (An analogous result holds for addition.) This allows us to regard $\mathbb{R}$ as a (proper) subset of $\mathbb{C}$:

$$\mathbb{R} \subset \mathbb{C}.$$

An arbitrary element is often denoted by the single letter z: $z = (x, y)$. The particular member of $\mathbb{C}$, $(0, 1)$, is everywhere denoted (except by electrical engineers) by the symbol i. Thus, by definition,

$$i = (0, 1).$$

Observe:

$$i^2 = i\cdot i$$
$$= (0,\ 1)\cdot(0,\ 1)$$
$$= (0\cdot 0 - 1\cdot 1,\ 1\cdot 0 + 0\cdot 1)$$
$$= (-1,\ 0).$$

But our identification gives $(-1, 0) = -1$. Hence,

$$i^2 = -1.$$

Thus the complex numbers $\mathbb{C}$ contain an element whose square equals -1.

One can now state and prove a theorem for $\mathbb{C}$ that is formally identical to theorem 2, parts A and M. In this case the additive identity will be $0 = (0, 0)$ and the multiplication identity $1 = (1, 0)$. The additive inverse of (x, y) is simply $(-x, -y)$. The multiplicative inverse of $(x, y) \neq (0, 0)$ turns out to be

$$\left(\frac{x}{x^2 + y^2},\ \frac{-y}{x^2 + y^2} \right).$$

(The proofs are straightforward and are left as exercises.)

Complex numbers may be more easily manipulated through a change of notation. Let $z = (x, y)$ be an arbitrary element of $\mathbb{C}$. Write

$$z = (x, y) = x + iy.$$

The new notation *allows complex numbers to be manipulated in a purely formal manner—using the commutative, associative, and distributive properties—as long as* i^2 *is replaced by* -1.

(Here's the proof for multiplication:

$$\left(x_1, y_1\right) \cdot \left(x_2, y_2\right) = \left(x_1 x_2 - y_1 y_2, \; x_2 y_1 + x_1 y_2\right)$$
$$= \left(x_1 x_2 - y_1 y_2\right) + i\left(x_2 y_1 + x_1 y_2\right).$$

Formally

$$\left(x_1 + iy_1\right) \cdot \left(x_2 + iy_2\right) = x_1 x_2 + x_1 iy_2 + iy_1 x_2 + i^2 y_1 y_2$$
$$= \left(x_1 x_2 - y_1 y_2\right) + i\left(x_2 y_1 + x_1 y_2\right).$$

The results are identical.)

As an example, let $z_1 = 2 + i$, $z_2 = 3 + 6i$. Then

$$z_1 + z_2 = \left(2 + i\right) + \left(3 + 6i\right)$$
$$= 5 + 7i$$

and

$$z_1 z_2 = \left(2 + i\right)\left(3 + 6i\right)$$
$$= 2 \cdot 3 + 2 \cdot 6i + i \cdot 3 + 6i^2$$
$$= 6 - 6 + 15i$$
$$= 0 + 15i$$
$$= 15i.$$

Complex numbers of the form $z = 0 + iy = iy$ are called *imaginary numbers*. The terminology goes back to René Descartes (1596–1650), who called $i = \sqrt{-1}$ "imaginary." Descartes' concern over the existence of the number i was shared by many other earlier mathematicians and it was not until the nineteenth century that complex numbers were legitimized and rigorously established. But the terminology persists and numbers such as $15i$ continue to be called "imaginary."

In a mathematical sense no number is more or less imaginary than another. All are ideas, mere pieces of imagination.

In summary, then, the complex numbers $\mathbb{C}$ consist of elements of the form $z = x + iy$ when $x, y \in \mathbb{R}$ and $i^2 = -1$. These numbers may be added and multiplied in a formal manner provided i^2, whenever it appears, is replaced by -1. A theorem formally identical to theorem 2, parts A and M, holds for $\mathbb{C}$. (Part O does not hold for $\mathbb{C}$. The complex numbers are not ordered. It makes no sense to ask whether or not $3 + 2i$ is larger or smaller than $6 - 4i$.)

The real numbers now become complex numbers of the form $z = x + i \cdot 0 = x$. And we have the completed chain of proper inclusions:

$$\mathbb{N} \subset \mathbb{Z} \subset \mathbb{Q} \subset \mathbb{R} \subset \mathbb{C}.$$

The complex numbers and the associated theory of functions of a complex variable became the glory of nineteenth-century mathematics. In that century mathematicians, having the stature of Gauss, Augustin Louis Cauchy (1789–1857), Karl Weierstrass (1815–1897), and Bernhard Riemann (1826–1866), turned their attention to the development of this mathematical theory.

Gauss defined complex integers (now called Gaussian integers). Cauchy proved the lovely integral theorem that bears his name. Weierstrass produced his theory of infinite series and brought full rigor to complex analysis.

In 1859 Riemann published a short paper that revealed deep connections between the new mathematics of complex numbers and Pythagoras's ancient subject of number theory. In this paper, Riemann wrote his famous conjecture on the location of the zeros of a complex function called the *zeta function*. This conjecture—now known as the *Riemann hypothesis*—stands today as the premier open problem in all of mathematics.

Near the end of the century Jacques Hadamard (1865–1963) and Charles-Jean de la Vallée Poussin (1866–1962)—independently of one another but both using methods of complex analysis—proved the prime number theorem. This theorem, which had challenged mathematicians for years, established the precise rate at which the prime numbers grow. Euclid told us in 300 BCE that the primes do not end and thus move out to infinity. Hademard and de la Vallée Poussin in 1896 found the exact pace at which they march.

Complex analysis flourishes today as a subject in its own right, and because of its applicability to other areas of mathematics—such as differential equations, number theory, Fourier series, and functions of several variables. The classical theory has not passed away. The work of the great masters still fascinates with its depth and beauty.

In 1748 Leonard Euler proved that for each real number x,

$$e^{ix} = \cos x + i \sin x$$

where the functions on the right-hand side are the ordinary trigonometric functions of cosine and sine. If we set $x = \pi$ we obtain

$$e^{i\pi} = \cos \pi + i \sin \pi.$$

But $\cos \pi = -1$ and $\sin \pi = 0$ (from trigonometry). Thus

$$e^{i\pi} = -1.$$

Hence

(e) $e^{i\pi} + 1 = 0.$

Once I heard a friend recite Tennyson's "Ulysses" in memory of a dead colleague. He spoke extemporaneously as we walked in gathering darkness on a college campus far from the ringing plains of windy Troy:

> The lights begin to twinkle from the rocks.
> The long day wanes; the slow moon climbs; the deep
> Moans round with many voices. Come, my friends.
> It is not too late to seek a newer world.[7]

John Gielgud could have said it no better. When he finished we walked in silence. Talking would have spoiled the beauty of the moment.

What can I say about equation (e) without spoiling its beauty? Let me just recite the facts: equation (e) contains

- the five most important numbers in mathematics: 1, 0, e, π, and i
- the most important relation: equality
- the three most important operations: addition, multiplication, and exponentiation.

Moreover the equation (e) contains nothing else, nothing extraneous. It is breathtakingly bare, like a poem by Robert Frost. When Euler saw the equation he wrote the word "startling." The contemporary mathematician, Herb Silverman, describes it as: "the most beautiful equation in all of mathematics."[8]

Equation (*e*) lives, of course, in the domain of complex analysis. Much, much beauty lives here. Elegance wanders by daylight through the entire mathematical world. But each night it comes home and sleeps in a house made of complex numbers.

Chapter 6

NUMBER MACHINES

The term "analysis" roughly describes part of the mathematical world that includes calculus and the higher subjects that proceed in the spirit of calculus. The real and the complex numbers constitute the soil from which these subjects rise. The subjects themselves—which perhaps range from classical differential equations on the one hand to modern functional analysis on the other—share an essence because each has been shaped through the common use of pieces of mathematical machinery called *functions*.

The concept of function can be precisely stated—as can most mathematical notions—in terms of sets. Analysts themselves, however, ordinarily think of functions as something more dynamic. They usually consider a function to be a machinelike device that transforms numbers in a certain set into numbers that belong to another set. We will also consider functions in this manner. And we will mostly restrict ourselves to situations in which the two sets are each sets of real numbers. The resulting functions then become what analysts call *real valued functions of real variables*.

DEFINITION 1. Let A and B be nonempty sets. Let f denote a rule that associates with each element of A a unique element of B. Then we write

$$f : A \to B$$

and say that f is a function from A into B. If f associates with $x \in A$ the element $y \in B$, we write

$$y = f(x).$$

We call $f(x)$ the value of f at x. (The symbol $f(x)$ is read "eff of x.") We refer to x as the independent variable and y as the dependent variable. The set A is called the domain of f. We write

$$D(f) = A$$

and say f is defined on A. The set $\{y : y = f(x), x \in A\}$ is called the range of f and we write

$$R(f) = \{y : y = f(x), x \in D(f)\}.$$

Figure 37 shows a general situation in which A and B are arbitrary sets and the function f associates $y \in B$ with a fixed $x \in A$. The domain of f is (by definition) the entire set A. The shaded subset of B denotes the range of f and illustrates the fact that $R(f)$ may not be all of B.

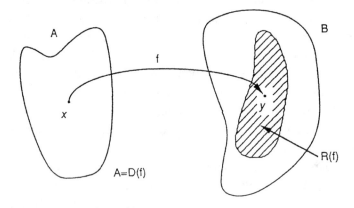

Figure 37: $f : A \rightarrow B$

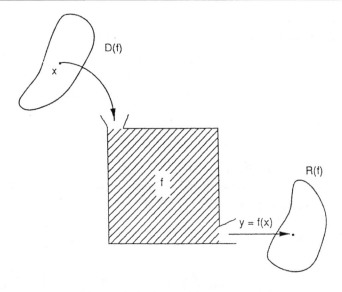

Figure 38: A machine named f

Figure 38 provides a mechanical illustration of a function as a striped box named f. You put x in at one end and $y = f(x)$ pops out at the other. The set from which the x's are selected constitutes the domain of f. The set of y's produced by the machine makes up the range of f.

Consider two examples:

EXAMPLE 1. Let $A = B = \mathbb{R}$ and let $f(x) = \pm x^2$. Here, in particular, $f(2) = \pm 2^2$ $= \pm 4$. Consequently, *the rule* f *does not define a function on* $\mathbb{R}$ because it fails to associate a unique number with each number in $\mathbb{R}$. ($f(2)$ tries to be two different values: 4 and –4.)

EXAMPLE 2. Let $A = B = \mathbb{R}$ and let $f(x) = x^2$. Here, f associates with each $x \in \mathbb{R}$ a *unique* $y \in \mathbb{R}$ given by $y = f(x) = x^2$. Thus f is a function from $\mathbb{R}$ into $\mathbb{R}$ and we may write $f : \mathbb{R} \to \mathbb{R}$. By definition of this function,

$$D(f) = \mathbb{R}.$$

But

$$R(f) = \left\{ y : y = x^2, \, x \in \mathbb{R} \right\}$$

and we know by theorem 3 of chapter 6 that $x^2 \geq 0$ for $x \in \mathbb{R}$. Thus $R(f) \neq \mathbb{R}$. (Actually, $R(f) = \{y : y \in \mathbb{R}, y \geq 0\}$.)

Notice that, in example 2, we have $f(2) = 4$ and $f(-2) = 4$. (In fact, $f(-x)$ $= x^2 = f(x)$ for all $x \in \mathbb{R}$.) Thus a function can transform two (or more) distinct values of $x \in A$ into the same value $y \in B$. What it cannot do (by the uniqueness part of the definition) is transform a single $x \in A$ into two (or more) distinct members of B.

But we want to be able to identify situations in which this does not happen, that is, situations in which f does not send two distinct x's to the same y. And we want also to distinguish cases in which $R(f) = B$. This leads to:

DEFINITION 2. Let $f : A \to B$. (Thus f is a function from A into B.)
(a) We say that f is *onto* B if and only if

$$R(f) = B.$$

(b) We say that $f : A \to B$ is one to one if and only if

$$x_1, x_2 \in A \text{ and } f(x_1) = f(x_2) \Rightarrow x_1 = x_2.$$

So f is *onto B* provided its range completely fills the set B. To say that $f : A \to B$ is one to one means that no two distinct members of A are transformed by f to the same member of B. The function $f : \mathbb{R} \to \mathbb{R}$ of example 2 fails to be onto $\mathbb{R}$ because (in particular) there is no $x \in \mathbb{R}$ such that $f(x) = x^2 = -1$. Moreover, this function is not one to one because (in particular) $f(-2) = 4 = f(2)$ but $2 \neq -2$.

Suppose, however, we modify the function of example 2 by changing the set B into $\{y : y \geq 0\}$.

EXAMPLE 3. Let $g : \mathbb{R} \to \{y : y \geq 0\}$ be defined by

$$g(x) = x^2$$

for each $x \in \mathbb{R}$. This function is an onto function because

$$R(g) = \{y : y \geq 0\}.$$

The function fails to be one to one because, again, $g(2) = 4 = g(-2)$ and $2 \neq -2$.

You may have noticed the close relationship between definition 2 and the chapter 2 notion of one-to-one correspondence. In fact, to say that there exists a one-to-one correspondence between sets S and T means—in the language of definition 2—that there exists a function

$$f : S \to T$$

such that f is both one to one and onto.

For example, the function $f : \mathbb{N} \to E$, defined by $f(n) = 2n$ for each $n \in \mathbb{N}$, is a one to one and onto function between the set $\mathbb{N}$ of natural numbers and the set E of positive even numbers. ($f : \mathbb{N} \to E$ is onto because if $y \in E$ then $y = 2m$ for some $m \in \mathbb{N}$. But $f(m) = 2m = y$. So the range of f fills all of E. f is one to one because, if $n, m \in \mathbb{N}$ and $f(n) = f(m)$, then $2n = 2m$. Then $n = m$.) Consequently, f produces a one-to-one correspondence between $\mathbb{N}$ and E.

DESCARTES' DREAM

Rene Descartes (1596–1650) receives the winner's share of credit for the invention of analytic geometry even though Fermat independently, and almost simultaneously, produced similar ideas. Before this work appeared, geometric questions about curves and other plane figures could be examined only in the light of elementary geometric methods that essentially had remained unchanged since Euclid's day. Analytic geometry brought revolutionary new techniques to bear on these problems and changed forevermore the manner in which mathematicians view geometry. E. T. Bell says of the importance of the new approach: "Descartes did not revise geometry; he created it."[1]

Legend tells us that Descartes' great idea came to him in a dream. If so, then it is good that no person from Porlock interrupted him before the dream was written. For, in a sense, the introduction of Descartes' methods marked the beginning of modern mathematics. Descartes believed he had found a "magic key." Perhaps so. But like many great ideas, this one was elegantly simple: *replace curves with functions so that geometry becomes analysis.*

Consider two copies of the real number line that intersect at right angles as shown in figure 39. Traditionally, the horizontal line is called the *x axis* and the vertical line the *y axis*. The arrows of figure 39 indicate the positive direction on each axis. The axes intersect at $x = 0$ and $y = 0$. The point of intersection is known as the *origin*. The axes divide the plane of figure 39 into four areas called *quadrants*. The quadrants are marked in counterclockwise order, I, II, III, IV.

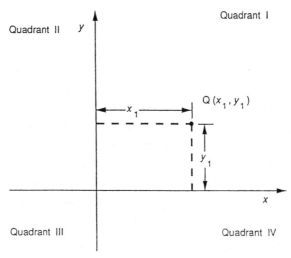

Figure 39: The Cartesian plane

Select an arbitrary point Q in the first quadrant. Determine the distance from Q to the y axis. Call this distance x_1. Now determine the distance y_1 from Q to the x axis. (See figure 39.) Next, *assign* to Q the pair of real numbers (x_1, y_1). These two numbers are known as the *Cartesian coordinates* of the point Q. (The Latinized version of "Descartes" is Cartesius.)

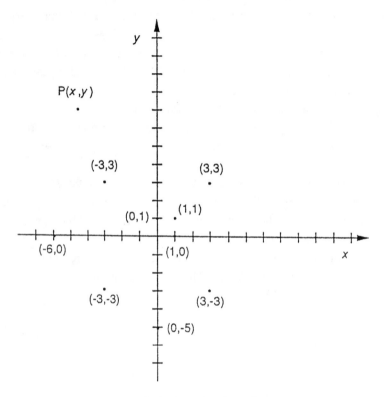

Figure 40: Points in the plane

Usually we denote an arbitrary point by P and its coordinates by (x, y). Figure 40 shows such a point, this time selected from quadrant II. In this case y is again chosen as the distance from P to the x axis but x is taken as the *negative of the distance* to the y axis. If P were selected from quadrant III, both x and y would be taken as negatives of the corresponding distances. We take y as negative and x as nonnegative when P lives in quadrant IV.

Thus if P has Cartesian coordinates (x, y) then

$$x \geq 0, y \geq 0 \text{ in quadrant I,}$$

$x \leq 0, y \geq 0$ in quadrant II,

$x \leq 0, y \leq 0$ in quadrant III,

and

$x \geq 0, y \leq 0$ in quadrant IV.

Figure 40 shows a selection of "plotted" points and their corresponding coordinates. Notice that points with $x = 0$ live on the y axis while those with $y = 0$ belong to the x axis. The origin is the point with coordinates $(0, 0)$.

As we have often done, we will identify the point with coordinates $(x, 0)$, with the real number x whenever such an identification is convenient. Similarly, the point $(0, y)$ will be identified with y.

Now we are ready for Descartes' magical idea. It is this:

Consider a curve C in the Cartesian plane. Let $P(x, y)$ be an arbitrary point on C. (See figure 41.) As P describes C, the coordinates x and y will vary in some manner. Perhaps these coordinates will satisfy an equation of the form $y = f(x)$ where f is a function. If this happens, then the geometric properties of the curve C will be reflected in the analytic properties of the function f. In order to know C, it is sufficient to know f. Geometry becomes analysis.

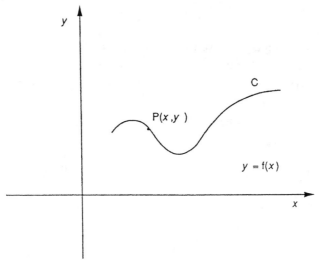

Figure 41: A curve described by a function

This idea revolutionized mathematics as relativity theory would later revolutionize physics. And Descartes knew what he had done: "he had surpassed all geometry before him as Cicero's rhetoric surpasses the ABC's."[2]

GRAPHS

DEFINITION 3. Let $f : A \to B$ be a function where A and B are sets of real numbers. The set of points in the Cartesian plane given by

$$G(f) = \{(x, y): x \in A, y = f(x)\}$$

is called the graph of f.

Recall that $A = D(f)$, the domain of f, so that

$$G(f) = \{(x, y): x \in D(f), y = f(x)\}.$$

Ordinarily, the graph of a function is a curve in the Cartesian plane. Examples follow:

EXAMPLE 4. Let $f : \mathbb{R} \to \mathbb{R}$ be given by

$$f(x) = 6 \text{ for all } x \in \mathbb{R}.$$

Hence, the graph is: $G(f) = \{(x, y): x \in \mathbb{R}, y = 6\}$. Hence

$$G(f) = \{(x, 6): x \in \mathbb{R}\}.$$

Thus the graph consists of those points in the Cartesian plane whose y coordinate equals 6, that is, those points whose distance from the y axis equals 6. As x varies of $\mathbb{R}$ these points describe the line parallel to the x axis and 6 units above it. (See figure 42.)

This is an example of a constant function. Each constant function has as its graph a line parallel to the x axis.

EXAMPLE 5. Let $f : \mathbb{R} \to \mathbb{R}$ be given by $f(x) = x$ for all $x \in \mathbb{R}$.

The graph is $G(f) = \{(x, y): x \in \mathbb{R}, y = x\}$. Hence

$$G(f) = \{(x, x): x \in \mathbb{R}\}.$$

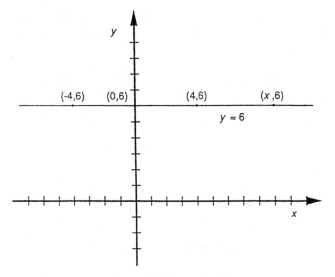

Figure 42: The graph of y = 6

Thus the graph consists of all those points in the Cartesian plane whose x and y coordinates are identical. (For example, $(1, 1)$, $(2, 2)$, $(-6, -6)$, $\left(-\dfrac{1}{2}, -\dfrac{1}{2}\right)$, $(\sqrt{2}, \sqrt{2})$.) The totality of these points generates the "45 degree line" shown in figure 43.

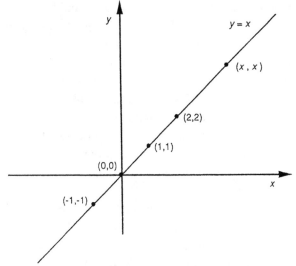

Figure 43: The graph of y = x

The preceding examples are special cases of the more general function $f(x) = ax + b$. It can be shown that the graph of such a function is always a straight line.

EXAMPLE 6. Let $f : \mathbb{R} \to \mathbb{R}$ be given by $f(x) = x^2$ for all $x \in \mathbb{R}$. Here, $G(f) = \{(x, y) : x \in \mathbb{R} \text{ and } y = x^2\}$. In this case, it is helpful to "plot some points" in order to determine the nature of the graph. Notice

$$x = 0 \Rightarrow f(x) = f(0) = 0^2 = 0$$

$$x = \frac{1}{2} \Rightarrow f(x) = f\left(\frac{1}{2}\right) = \left(\frac{1}{2}\right)^2 = \frac{1}{4}$$

$$x = 1 \Rightarrow f(x) = f(1) = 1^2 = 1$$

$$x = 2 \Rightarrow f(x) = f(2) = 2^2 = 4$$

$$x = 3 \Rightarrow f(x) = f(3) = 3^2 = 9.$$

Thus the points $(0, 0)$, $\left(\frac{1}{2}, \frac{1}{4}\right)$, $(1, 1)$, $(2, 4)$, and $(3, 9)$ are all on the graph of $y = x^2$.

Notice also that, for this particular function

$$f(-x) = (-x)^2 = x^2 = f(x).$$

Thus (x, y) belongs to the graph of $y = f(x)$ if and only if the graph also passes through the point $(-x, y)$. But, since the points $(-x, y)$ and (x, y) are symmetrically placed with respect to the y axis, this means that the graph of $y = x^2$ is symmetric with respect to the y axis.

With this information, and the particular points given above, we see that the graph of $y = x^2$ has the form shown in figure 44.

The graph of $y = x^2$ represents a particular case of a collection of curves known as a *parabola*. It is possible to prove that the graph of any function of the form $f(x) = ax^2 + bx + c$ is a parabola provided $a \neq 0$. (The parabola opens upward if $a > 0$ and downward if $a < 0$. Of course, if $a = 0$ then $f(x) = bx + c$, whose graph is a straight line.)

Figure 45 shows the downward-turning parabolic graph of $f(x) = -x^2 + 2x + 3$. You can easily verify the shape of the graph by plotting a few points. (A better—albeit trickier—method is to notice that $f(x) = -(x - 1)^2 + 4$ and then to argue that the graph of $y = f(x)$ is merely *an appropriately shifted graph of $y = -x^2$.*)

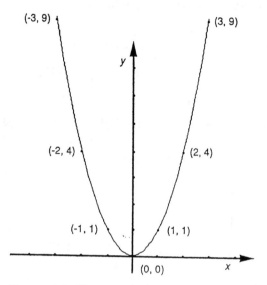

Figure 44: The parabolic graph of $y = x^2$

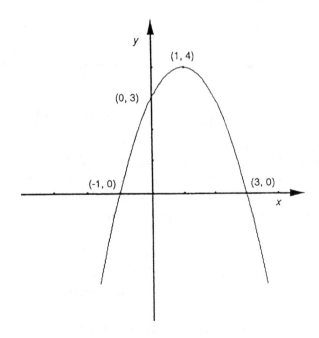

Figure 45: The parabolic graph of $y = -x^2 + 2x + 3$

DISTANCE BETWEEN POINTS

The graph of the function $f : A \rightarrow B$, being the set

$$G(f) = \{(x, y) : x \in A, \, y = f(x)\},$$

is thus identical to the *graph of the equation* $y = f(x)$. But not every equation defines a function and hence the graph of an equation may not represent the graph of a function. In particular, *it is false that each curve in the Cartesian plane represents the graph of a function.*

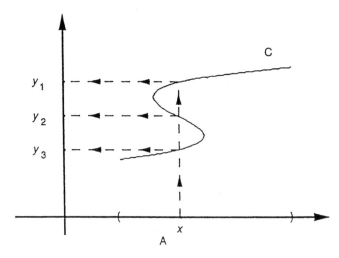

Figure 46: A curve that is not the graph of a function

Figure 46, for example, presents a smooth curve C in the Cartesian plane, winding over an interval A of real numbers. But C cannot be the graph of a function defined on A because, as figure 44 shows, the curve forces three y values to correspond to the indicated value of x in A. This violates the definition of the function concept, which asserts the correspondence of a unique value of y to each x via the equation $y = f(x)$.

If C is to be the graph of a function $f : A \rightarrow B$ then, when we start with $x \in A$, the curve must select for us a unique $y \in B$, as we go vertically from x up to the curve and then horizontally over to B. In order for a curve to represent the graph of a function, it must be true that each line, perpendicular to the x axis, intersects the curve in at most one point. Such a curve—one that is the graph of a function—appears in figure 47.

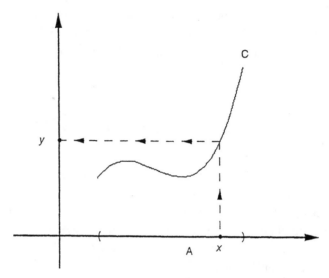

Figure 47: A curve that is the graph of a function

This discussion shows that an ordinary circle cannot be the graph of a function since certain vertical lines cut a given circle in two points. However, any circle represents the graph of a well-defined equation. And—since a circle can be described as the collection of points at a fixed distance from a given point—we can determine this equation once we have an analytic expression for distance.

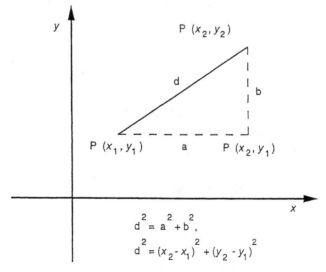

$$d^2 = a^2 + b^2,$$
$$d^2 = (x_2 - x_1)^2 + (y_2 - y_1)^2$$

Figure 48: The distance formula

Consider two distinct points P_1 and P_2 having coordinates (x_1, y_1) and (x_2, y_2), respectively. Let $d = d(P_1, P_2)$ denote the distance between P_1 and P_2. We see from the illustration in figure 46 and an application of the Pythagorean theorem that

$$d^2 = a^2 + b^2$$
$$= \left(x_2 - x_1\right)^2 + \left(y_2 - y_1\right)^2.$$

Thus

$$d = \sqrt{\left(x_2 - x_1\right)^2 + \left(y_2 - y_1\right)^2}.$$

(Although figure 46 shows the situation $x_2 > x_1$ and $y_2 > y_1$, the argument holds in general since $x_2 - x_1$ and $y_2 - y_1$ will always represent the appropriate lengths a and b or the negative of these lengths. So $(x_2 - x_1)^2 = a^2$ and $(y_2 - y_1)^2 = b^2$ in either case.)

We have proved *the distance formula*: Let $P_1 (x_1, y_1)$ and $P_2(x_2, y_2)$ be any two points in the Cartesian plane. Then the distance between P_1 and P_2 is given by:

$$d(P_1, P_2) = \sqrt{\left(x_2 - x_1\right)^2 + \left(y_2 - y_1\right)^2}.$$

Now consider the circle C with center at $P_0 (x_0, y_0)$ and radius r. Let $P(x, y)$ be an arbitrary point. Then P is on the circle C (see figure 49) if and only if $d(P, P_0) = r$. Since these are nonnegative quantities, this holds if and only if

$$\left(d\left(P, P_0\right)\right)^2 = r^2$$

or

(c)
$$\left(x - x_0\right)^2 + \left(y - y_0\right)^2 = r^2$$

Thus equation (c) is the equation of circle C.

Euclid saw a circle as a closed curve like that of figure 49. To Descartes, and the mathematicians who followed him, a circle became indistinguishable from the mathematical equation that described it.

Consider the special case of the *unit circle*: $x_0 = y_0 = 0$ and $r = 1$. Its equation is

(d)
$$x^2 + y^2 = 1.$$

Figure 50 shows the unit circle and the line

(*e*) $y = x$

of figure 43.

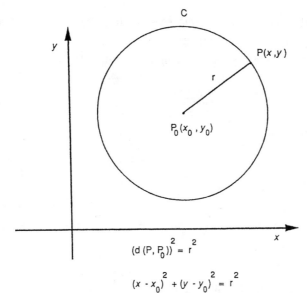

$$(d\,(P, P_0))^2 = r^2$$

$$(x - x_0)^2 + (y - y_0)^2 = r^2$$

Figure 49: Equation of a circle

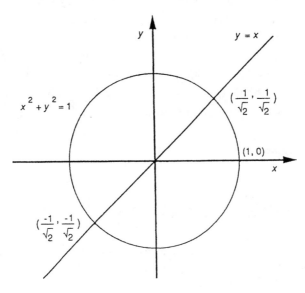

Figure 50: The unit circle and the line $y = x$

We might be hard-pressed to find the exact intersection of these two geometrical figures using only the methods of Euclidean geometry. But Descartes' magic key makes it easy. A point will belong to both the circle and the line, and thus be a point of intersection, if and only if its coordinates simultaneously satisfy both of equations (d) and (e).

Equation (e) says $y = x$. Substituting this value into (d) gives:

$$x^2 + x^2 = 1 \Rightarrow 2x^2 = 1$$

$$\Rightarrow x^2 = \frac{1}{2}$$

$$\Rightarrow x = \pm \frac{1}{\sqrt{2}}.$$

Using $y = x$ again we obtain $x = \dfrac{1}{\sqrt{2}}$ and $y = -\dfrac{1}{\sqrt{2}}$, or $x = -\dfrac{1}{\sqrt{2}}$ and $y = -\dfrac{1}{\sqrt{2}}$.

Thus the only candidates for intersection are $\left(\dfrac{1}{\sqrt{2}}, \dfrac{1}{\sqrt{2}} \right)$ and $\left(\dfrac{-1}{\sqrt{2}}, \dfrac{-1}{\sqrt{2}} \right)$.

A quick check shows these coordinates actually satisfy both equations so they indeed are the exact points of intersection. (We need to make the check since the above implications go only one way.)

In this case the calculations of the intersection points were routine and easy. Nevertheless, this computation illustrates the importance and the power of Descartes' methods. In theory you can find the points of intersection of any two curves by determining their equations and solving them simultaneously. The curves could be simple like our circle and line, or they might be complicated, like the path of a space shuttle and the orbit of Venus.

COMBINATIONS OF FUNCTIONS

We can build complicated functions from simple ones by combining them in various ways. The standard combining methods are *addition*, *multiplication*, *division*, and *composition*:

DEFINITION 3. Let f and g be functions with common domains of defini-

tion. We define $f + g, f \cdot g$, and $\dfrac{f}{g}$ by

$$(f + g)(x) = f(x) + g(x),$$
$$(f \cdot g)(x) = f(x) \cdot g(x), \text{ and}$$
$$\left(\frac{f}{g}\right)(x) = \frac{f(x)}{g(x)}, \text{ if } g(x) \neq 0.$$

Thus, for example, if $f(x) = x^2$ and $g(x) = 2x + 1$ for $x \in \mathbb{R}$, then

$$(f + g)(x) = x^2 + 2x + 1 = (x + 1)^2,$$
$$(f \cdot g)(x) = x^2(2x + 1) = 2x^3 + x^2, \text{ and}$$
$$\left(\frac{f}{g}\right)(x) = \frac{x^2}{2x + 1}.$$

(Notice that $\dfrac{f}{g}$ is not defined when $2x + 1 = 0$, i.e., $x = -\dfrac{1}{2}$.)

DEFINITION 4. Let f and g be functions with $R(f) \subset D(g)$. (g is defined on the range of f.) Then the composition of f and g, denoted by $f \circ g$, is defined by

$$(f \circ g)(x) = f(g(x)).$$

As an example, again let $f(x) = x^2$ and $g(x) = 2x + 1$. Since g is defined for all $x \in \mathbb{R}$, its domain includes the range of f. Then

$$(f \circ g)(x) = f(g(x))$$
$$= (g(x))^2$$
$$= (2x + 1)^2$$
$$= 4x^2 + 4x + 1.$$

But

$$(g \circ f)(x) = g(f(x))$$
$$= 2f(x) + 1$$
$$= 2x^2 + 1.$$

So, in general, $f \circ g \neq g \circ f$. *Thus the composition of functions is not commutative.*

INVERSE FUNCTIONS

Suppose $f : A \rightarrow B$ is one to one and onto. This means that the equation $y = f(x)$ associates with each $x \in A$ exactly one $y \in B$ and conversely. Thus we can *start* with a $y \in B$ and then select the unique $x \in A$ such that $y = f(x)$. (There exists such an x because f is onto. There is only one such x because f is one to one.)

This process enables us to *define* a function $g : B \rightarrow A$ by writing

$$x = g(y) \Leftrightarrow y = f(x).$$

The function g is called the *inverse* of f and we write $g = f^{-1}$.

EXAMPLE 7. Let $f : \mathbb{R} \rightarrow \mathbb{R}$ be defined by $f(x) = 2x + 1$. Then f is one to one and onto. f is one to one because $f(x_1) = f(x_2) \Rightarrow 2x_1 + 1 = 2x_2 + 1 \Rightarrow x_1 = x_2$. f is onto because, if $y \in \mathbb{R}$ then

$$\frac{y-1}{2} \in \mathbb{R} \text{ and } f\left(\frac{y-1}{2}\right) = 2\left(\frac{y-1}{2}\right) + 1 = y - 1 + 1 = y.$$

Thus f has an inverse function g defined by

$$x = g(y) \Leftrightarrow y = f(x)$$
$$\Leftrightarrow y = 2x + 1$$
$$\Leftrightarrow x = \frac{y-1}{2}.$$

Hence, $g(y) = f^{-1}(y) = \frac{y-1}{2}$.

Notice that we obtain the inverse function here by simply solving the equation $y = f(x)$ for x in terms of y. This procedure is standard and when it fails the failure *often* tells you that f^{-1} does not exist. For example, if $f(x) = x^2$. Then

$$y = f(x) \Rightarrow y = x^2$$

$$\Rightarrow x = \sqrt{y} \text{ or } x = -\sqrt{y}.$$

Thus there are at best two different solutions for x in terms of y and, at worst, no solution at all (when $y < 0$). But something like this might have been expected since, considered as a function from $\mathbb{R}$ into $\mathbb{R}$, $f(x) = x^2$ is neither one to one nor onto. So f^{-1} does not exist. (However, if we consider $h : \{x : x \geq 0\} \to \{y : y \geq 0\}$ defined by $h(x) = x^2$ then h is one to one and onto and $h^{-1}(y) = \sqrt{y}$.)

In general, if f is one to one and onto then

$$y = f(x) \Leftrightarrow x = f^{-1}(y).$$

Hence,

$$(f \circ f^{-1})(y) = f(f^{-1}(y))$$

$$= f(x)$$

$$= y.$$

Thus

$$(f \circ f^{-1})(y) = y, \; y \in B.$$

Similarly,

$$(f^{-1} \circ f)(x) = x, \; x \in A.$$

Mathematicians often summarize this situation by saying $f \circ f^{-1}$ is the identity function on B and $f^{-1} \circ f$ is the identity function on A.

TRANSCENDENTAL NUMBERS

The operations of combination allow us to form new functions from known functions. For instance, if we start with the identity function on $\mathbb{R}$

$$f(x) = x$$

then we can, by appropriate multiplication and addition, form functions like

$$p_1(x) = 2x + 1,$$

$$p_2(x) = 3x^2 + 4x + \sqrt{3}, \text{ and}$$

$$p_3(x) = 16x^3 - \frac{2}{3}x^2 + \pi x + 1.$$

Functions of these types are called *polynomials*. The *general polynomial of degree n* is defined by

(f) $$p(x) = a_n x^n + a_{n_1} x^{n-1} + \dots + a_1 x + a_0.$$

where n is a natural number or zero and each a is a real number.

We looked at the general polynomial of degree 2,

$$g(x) = ax^2 + bx + c$$

when we discussed the quadratic formula earlier. Then we saw that $q(x) = 0$ had a real-number solution (given by the quadratic formula) provided $b^2 - 4ac \geq 0$.

Much of mathematical analysis deals with properties of polynomials, often with *the zeros of polynomials*, which are the values of x for which $p(x) = 0$. A famous theorem, *the fundamental theorem of algebra*, asserts that *each polynomial given by equation* (f) *has a zero. The zeros, however, may not be real numbers.* But they can always be found in $\mathbb{C}$, the larger set of complex numbers. The fact that these roots always exist in $\mathbb{C}$—but not in $\mathbb{R}$—provides a fundamental reason for developing the complex numbers in the first place.

Of course, once you delve into these complex numbers, you turn quickly to polynomials more general than those given by (f). In $\mathbb{C}$ you study polynomials whose coefficients (the a_k's) are themselves complex numbers. Such matters are far beyond the scope of this book. But we can conclude our introduction to functions by examining a connection between analysis and number theory that involves polynomials that are *less general* than those given by equation (f).

DEFINITION 5. A real number w is called algebraic if w is a zero of some polynomial

$$p(x) = b_n x^n + b_{n-1} x^{n-1} + \ldots + b_0$$

where $b_k \in \mathbb{Z}$, $k = 0, 1, \ldots, n$.

(Thus w is algebraic if $p(w) = 0$ when p is some polynomial with integer coefficients.)

Examples of algebraic numbers are $\frac{1}{2}$ and $-\frac{2}{3}$ because $2x - 1 = 0$ when $x = \frac{1}{2}$ and $3x + 2 = 0$ when $x = -\frac{2}{3}$.

Similarly, any rational number is algebraic because $w = \frac{p}{q}$ is a zero of the polynomial

$$qx - p = 0.$$

But certain irrational numbers are also algebraic. For example, $\sqrt{2}$ is algebraic because it is a zero of

$$x^2 - 2 = 0.$$

Similarly, $\sqrt{3}, \sqrt{5}, \sqrt{7}, \ldots$ are all algebraic numbers.

However, it can be shown (by a not-too-difficult argument that we shall omit) that the *set of all algebraic numbers is a countable set*. Since the real numbers are not countable, *it follows that there exists an uncountable infinity of nonalgebraic numbers. Such numbers are called transcendental.* Euler first considered these numbers around 1744.

Even though the real numbers contain an uncountable infinity of transcendental numbers, it is usually difficult to prove that any particular number is transcendental. The best-known examples of transcendatal numbers are e and π. Ferdinand Linderman (1852–1939) showed that π is transcendental in 1882. Charles Hemite (1822–1901) established the transcendental character of e in 1873. Joseph Liouville (1809–1882) produced the first specific examples of transcendental numbers in 1844. A full century had elapsed between the time Euler defined the notion of a transcendental number and the appearance of Liouville's example.

Transcendental numbers are ubiquitous because they are uncountable. But finding any particular one is not easy. It's like looking for a light switch in a dark hotel room. You know there are plenty around. But you need the lights on first to find one.

Chapter 7

PROBABILITY

The Romantic poets disdained both the deterministic physics of Isaac Newton and the mathematical arguments that produced it. John Keats doubted that truth could ever be reached by consecutive reasoning. Earlier, in the "Dunciad," Alexander Pope worried about an "all-composing" science that might lead to

Art after Art goes out, and all is night.[1]

And the mystical Blake warned against:

"In here/out there" logic,[2]

and in a painting he depicted Newton as a man lost in concentration, oblivious to all nature, hunched and naked on the ocean's floor.

Newton's mathematics yielded a completely predictable universe in the sense that all motion—whether the whirling of stars in the heavens or the movement of whales in the deep—is determinable. If you know where the thing is now and which forces are acting on it, you can compute its future position. Keats and the others found this disconcerting or irrelevant.

But two great discoveries have recently altered our view of the real world. One of these discoveries is Einstein's theory of relativity; the other is quantum mechanics. Einstein's theory dramatically altered our conceptions of space, time, and the geometry of the universe. But, revolutionary as it is, it preserved the deterministic aspects of Newtonian mechanics. Like Newton's world, Einstein's universe is predictable. The Romantics would not have been pleased with a relativistic universe. Blake would cast Einstein also into the sea.

Quantum theory, however, is another matter. Determinism disappears from the quantum world. The occupants of this world, subatomic particles, move erratically—first here, then there, with no continuous motion in between. A key element of this theory—the *Heisenberg uncertainty principle*—tells us that we cannot ever know simultaneously both the position and the state of motion of a particular particle.

Calculus and differential equations ruled Newton's world. Einstein's universe reflected the non-Euclidean geometry of Riemann. But the quantum world is governed by the mathematics of chance and uncertainty. This mathematics is called *probability theory*. In the probabilistic world, randomness dominates. Nothing is certain here but uncertainty. What, I wonder, would Blake make of this?

FINITE SAMPLE SPACES

We begin with an example.

EXAMPLE 1. Toss a fair coin three times. What is the probability that the coin lands heads up exactly twice?

SOLUTION. First of all, we shall refer to the coin-tossing process as an *experiment*. Our experiment, then, consists of three stages, each stage being a single toss of the coin.

Next, we need to determine the set S of all possible outcomes of the experiment. Such a set is called the *sample space* for the experiment. A convenient method for determining the possible outcomes is described in figure 51 called a *tree diagram*.

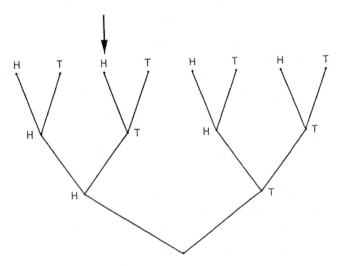

Figure 51: A tree diagram for three coin tosses

The first two branches from the bottom of the tree describe the possible outcomes of the first stage of the experiment. On the first toss, the coin lands heads or else it lands tails. These possibilities are indicated by H and T, respectively, marked at the end of the two branches.

The next four branches show the possibilities at the second stage—the second toss of the coin. The tree is continued in this manner until all possibilities have been sketched. Each path through the tree of the completed diagram represents a possible outcome of our experiment. (For example, the path of figure 51 marked with the arrow indicates the possible outcome H, T, H: heads on the first toss, tails on the second toss, and heads on the third toss.) Since there are eight different paths through the tree, our experiment has eight possible outcomes.

Next, we write the set S of possible outcomes in standard set-theoretic form. One way is:

$$S = \{(H, H, H), (H, H, T), (H, T, H), (H, T, T), (T, H, H), (T, H, T),$$
$$(T, T, H), (T, T, T)\}.$$

Thus each member of S is a triple of symbols like (H, T, H), which again represents the outcome: heads on the first toss, tails on the second toss, and heads on the third toss. S contains exactly eight elements so that, using the notation of chapter 2

$$n(S) = 8.$$

The set S is the sample space for the particular experiment consisting of three tosses of a coin.

Our original question concerned the probability of getting exactly two heads in three coin tosses. The elements of S that correspond to exactly two heads are (H, H, T), (H, T, H), and (T, H, H). The subset of S determined by these elements is

$$A = \{(H, H, T), (H, T, H), (T, H, H)\}.$$

A subset of a sample space is called an *event*. In this language, then, we are asked to determine "the probability of the event A." The common symbol for this is $P(A)$. Thus our question becomes: "What is $P(A)$?"

Finally, we use the given information that the coin we tossed three times is a *fair coin*. To say the coin is fair means—by definition—that any member of the sample space is as likely to occur as any other member. And this tells us—again by definition—that "the probability of event A equals the number of ways that A can occur divided by the total number of possible outcomes."

Hence

$$P(A) = \frac{n(A)}{n(S)}$$

or

$$P(A) = \frac{3}{8}.$$

Therefore, the probability of obtaining exactly two heads in three tosses of a fair coin is $\frac{3}{8}$.

Example 1 generalizes in a natural manner to extend the definition of the concept to finite sample space in which the possible outcomes are all equally likely. In this situation the probability notion bears the rather fancy title of *equiprobable probability measure*.

DEFINITION 1. Let S be a finite sample space. Let $A \subset S$. Define $P(A)$ by

$$P(A) = \frac{n(A)}{n(S)}.$$

Then P is known as the equiprobable probability measure and $P(A)$ is called the probability of A.

Notice that P is defined on subsets of S and that $0 \leq P(A) \leq 1$. ($n(A)$ cannot exceed $n(S)$ because A is a subset of S.) In particular,

$$P(\varnothing) = \frac{n(\varnothing)}{n(S)} = 0$$

and

$$P(S) = \frac{n(S)}{n(S)} = 1.$$

Once again subsets of S are called events so that "$P(A)$ equals the probability of event A." (We use the terms "event" and "sample space" to remind us of the connection with real-world experiments such as the tossing of coins. But S is merely a finite set and A a subset.)

EXAMPLE 2. A fair die is rolled once. What is the probability that an even number comes up? Solution: Here the experiment may be restated as: select a number at random from the numbers 1, 2, 3, 4, 5, 6.

The sample space is

$$S = \{1, 2, 3, 4, 5, 6\}.$$

Let A be the event:

$$A = \text{"an even number comes up."}$$

Then

$$A = \{2, 4, 6\}.$$

The phrase "at random" tells us to use the equiprobable probability measure. Hence

$$P(A) = \frac{n(A)}{n(S)}$$

$$= \frac{3}{6}$$

$$= \frac{1}{2}.$$

So far, matters are theoretically simple because we are dealing only with equiprobable measure. (But, as we shall see, applications can be devilishly complicated even in this situation.) The next definition extends the probability concept to cover cases where outcomes may not be equally likely—cases where we might toss biased coins, or roll loaded dice.

DEFINITION 2. Let S be a finite sample space:

$$S = \{x_1, x_2, \ldots, x_n\}.$$

Let w be a function defined on S with the property that

$$w(x_k) \geq 0 \text{ for each } x_k \in S$$

and

$$w(x_1) + w(x_2) + \ldots w(x_n) = 1.$$

If A is any nonempty subset of S:

$$A = \{y_1, y_2, \ldots, y_m\}.$$

Then we define the probability of A, $P(A)$, by

$$P(A) = w(y_1) + w(y_2) + \ldots + w(y_m).$$

We define $P(\emptyset) = 0$.

Then P is called a probability measure on S. (More properly, P is a probability measure on subsets of S. In the notation, $A = \{y_1, y_2, \ldots y_n\}$, each y_k is one of the x_js since $A \subset S$.)

Some comments on definition 2 are in order:

(1) The function w defined on S is called a *weight function*.

(2) The value $w(x_k)$ is called the *weight of the element* $x_k \in S$.

(3) The weight of any element of S is nonnegative and the sum of all the weights of the elements of the sample space equals 1.

(4) The probability, $P(A)$, of an event A is the sum of all the weights of elements of A.

The awkward notation for A and $P(A)$ in definition 2 can be avoided if we use summation notation. By definition

$$\sum_{k=1}^{m} t_k = t_1 + t_2 + \ldots + t_m.$$

This equation is read: "the sum from $k = 1$ to m of t_k equals t_1 plus t_2 plus . . . plus t_m."

For example,

$$\sum_{k=1}^{6} k^2 = 1^2 + 2^2 + 3^2 + 4^2 + 5^2 + 6^2$$

and

$$\sum_{k=1}^{n} k = 1 + 2 + 3 + 1 + 2 + 3 + \ldots + n.$$

The index *k*—the index of summation—is a dummy index in the sense that it can be replaced by any other appropriate index without changing the value of the summation. For example,

$$\sum_{s=1}^{6} s^2 = 1^2 + 2^2 + 3^3 + 4^2 + 5^2 + 6^2 = \sum_{k=1}^{6} k^2.$$

We may also use summation notation of the form

$$\sum_{x \in B} f(x) = \text{"sum of the values of } f(x)$$
$$\text{for those } x\text{'s that belong to the set } B.\text{"}$$

For example, if $B = \{-1, 2, 4\}$ then

$$\sum_{x \in B} x^2 = \sum_{x \in \{-1, 2, 4\}} x^2 = \left(-1^2\right) + 2^2 + 4^2 = 21.$$

Again, we see that the index *x* can be replaced by another:

$$\sum_{x \in B} f(x) = \sum_{y \in B} f(y).$$

Using this notation, the sums that appear in definition 2 become

$$\sum_{k=1}^{n} w(x_k) = 1$$

and

$$P(A) = \sum_{x \in A} w(x).$$

Suppose we are given a biased coin:

EXAMPLE 3. A particular biased coin is twice as likely to land heads as tails. The coin is tossed once. What is the probability that the coin lands heads?

Solution. Since the coin is not fair, the equiprobable measure (definition 1) does not apply. We must, therefore, determine the requested probability by defining a weight function on an appropriate sample space. Since the coin is tossed only once we may take as sample space

$$S = \{H, T\}.$$

At this stage we can, theoretically, assign weights to the two elements of S in infinitely many ways—subject only to the conditions that the weights are nonnegative and they sum to 1. But we are told that our coin is twice as likely to land heads as tails. We take this to mean that, if we call the weight function w:

$$w(H) = 2w(T).$$

That is, we define w so that the weight it assigns to H is twice what it gives to T.

Admittedly, this procedure is somewhat arbitrary. But there seems to be no other way to interpret the nonmathematical piece of information: "the coin is twice as likely to land heads as tails." We are, after all, building here a mathematical model of a physical coin. Our model may not be a perfect copy of a coin, just as Monet's *Waterloo Bridge* is not a perfect replica of that passage over the Thames. But we must have a model, inexact or not. Probability theory—like all of mathematics—deals with mathematical objects, not with silver coins.

Once it has been set down, however, the condition

$$w(H) = 2w(T)$$

completely determines the probability measure. We must have

$$\sum_{x \in s} w(x) = 1$$

or

$$w(H) + w(T) = 1.$$

Hence

$$2w(T) + w(T) = 1$$

so

$$w(T) = \frac{1}{3}.$$

Then

$$w(H) = 2w(T) = \frac{2}{3}.$$

So, this biased coin lands heads with probability $\frac{2}{3}$.

Notice that, in example 3 we properly write $w(H)$ but we write $P(\{H\})$. The weight function w is defined on S but the probability function P is defined on subsets of S. It is *incorrect* to write $P(H)$. However, we will often write symbols like, P ("the coin lands heads"), when we understand that the phrase in quotation marks *defines* a particular subset of S.

Let's also notice that the event B:

$$B = \text{"the coin lands tails"}$$

has probability

$$P(B) = P(\{T\}) = \frac{1}{3}.$$

Thus, $P(A) + P(B) = 1$. But here, $B = S \setminus A$, so that

$$P(A) + P(S \setminus A) = 1.$$

We will see shortly that this always happens—in every sample space, for each probability measure, P, whether equiprobable or not.

However, before we turn to this and other general properties of probability measures, let's determine the relationship between definition 1 and definition 2. Each defines something called a probability measure. We must show they are consistent.

THEOREM 1. Let S be a finite sample space,

$$S = \{x_1, x_2, \ldots, x_m\}.$$

Let w be defined on S by

$$w(x_k) = \frac{1}{n(S)} = \frac{1}{m}$$

for each $x_k \in S$. Then w is a weight function and let P be the resulting probability measure given by definition 2. Then P is the equiprobable probability measure of definition 1.

PROOF. First of all, w is a weight function because

$$w(x_k) = \frac{1}{m} \geq 0$$

and

$$\sum_{k=1}^{n} w(x_k) = w(x_1) + w(x_2) + \ldots + w(x_m)$$

$$= \frac{1}{m} + \frac{1}{m} + \ldots + \frac{1}{m}$$

$$= m \cdot \frac{1}{m}$$

$$= 1.$$

(The second equality from the bottom holds because the summation contains exactly m terms.)

Now let A be any subset of S. Then

$$P(A) = \sum_{x \in A} w(x)$$

$$= \sum_{x \in A} \left(\frac{1}{m} \right)$$

$$= \frac{1}{m} + \frac{1}{m} + \ldots + \frac{1}{m}.$$

This last sum contains a $w(x) = \frac{1}{m}$ for each $x \in A$. Thus the sum contains $n(A)$ terms. This gives

$$P(A) = n(A) \cdot \frac{1}{m}$$

$$= \frac{n(A)}{n(S)}$$

because $n(S) = m$.

Therefore, P is the equiprobable measure of definition 1.

Therefore, theorem 1 assures the consistency of the two definitions—definition 2 including definition 1 as a special case. We began with definition 1 because it seems more natural.

Certain situations, however, may be interpreted either way. You may think of them as "fair and equiprobable" or as "biased" (and thus requiring the use of definition 2), depending on the choice of sample space.

EXAMPLE 4. A box contains 3 red balls and 5 blue balls. A ball is selected at random from the box. What is the probability that a red ball is selected?

SOLUTION I. Since we are only interested in the color of the chosen ball, we may use as sample space

$$S = \{R, B\}.$$

We want to assign weights $w(R)$ and $w(B)$ to each element of S. Since the box contains 3 red balls and 5 blue balls, a natural choice is

$$w(R) = \frac{3}{5} w(B).$$

Then

$$w(R) + w(B) = 1$$

gives

$$\frac{3}{5} w(B) + w(B) = 1.$$

So,

$$\frac{8}{5} w(B) = 1$$

and

$$w(B) = \frac{5}{8}.$$

Then,

$$w(R) = \frac{3}{5}w(B) = \frac{3}{5} \cdot \frac{5}{8} = \frac{3}{8}.$$

Finally,

$$P \text{ ("red ball is selected")} = P(\{R\})$$

$$= \frac{3}{8}.$$

SOLUTION II. This time consider the balls separately and use the obvious sample space

$$S = \{r_1, r_2, r_3, b_1, b_2, b_3, b_4, b_5\}$$

where $n(S) = 8$. Since the ball is selected randomly, any one ball is as likely to be chosen as any other ball. Thus equiprobable measure applies. If A = "a red ball is selected" then

$$P(A) = P(\{r_1, r_2, r_3\})$$

$$= \frac{n(A)}{n(S)}$$

$$= \frac{3}{8}.$$

Of these two methods I prefer solution II since it used the more natural equiprobable measure. Nevertheless, solution I is instructive since it illustrates the two key steps in the solution of most probability problems: formulation of the sample space and determining the appropriate probability measure. Take these two steps properly and the answer will fall in your lap.

FUNDAMENTAL PROPERTIES

From definition 2 we can deduce several fundamental (and important) properties of an arbitrary probability measure. (Remember that an "event" is a subset of the sample space.)

THEOREM 2. Let S be a finite sample space and let P be a probability measure on S given by definition 2. Then

(1) $P(\emptyset) = 0$ and $P(S) = 1$,
(2) For any event A, $0 \leq P(A) \leq 1$,
(3) If A and B are events with $A \cap B = \emptyset$, then $P(A \cup B) = P(A) + P(B)$.

PROOF. (1) $P(\emptyset) = 0$ and $P(S) = \sum_{x \in S} w(x) = 1$ are requirements of definition 2 so there is nothing to prove. (They are included here only for emphasis.)

(2) By definition $P(A) = \sum_{x \in A} w(x)$ and $w(x) \geq 0$. Hence,

$$0 \leq P(A) = \sum_{x \in A} w(x) \leq \sum_{x \in S} w(x) = 1.$$

(The inequality follows because the right-hand summation contains all terms of the left-hand summation and probably more.) So, $0 \leq P(A) \leq 1$.

(3) Let $A \cap B = \emptyset$. Then A and B have no elements in common. Hence,

$$P(A \cup B) = \sum_{x \in A \cup B} w(x)$$
$$= \sum_{x \in A} w(x) + \sum_{x \in B} w(x)$$
$$= P(A) + P(B).$$

This completes the proof.

The condition $A \cap B = \emptyset$ of part (3) means that A and B are disjoint sets. However, in the context of probability theory, this property carries a different name.

DEFINITION 3. Let A and B be subsets of a finite sample space S. We say that A and B are mutually exclusive events if $A \cap B = \emptyset$.

Also in the context of probability theory—we often read $A \cup B$ as "either A or B." Thus—in this language—part (3) of theorem 2 says:

If A and B are mutually exclusive events, then the probability of A or B equals the probability of A plus the probability of B.

Similarly, we may refer to $A \cap B$ as "A and B." An important notion associated with $A \cap B$ is the concept of *independence*.

DEFINITION 4. Let A and B be subsets of a finite sample space. We say that A and B are independent events if $P(A \cap B) = P(A) \cdot P(B)$.

In any particular example, we often assume mutual exclusiveness or independence directly from the real-world situation. Such assumptions, of course, then become assumptions about the mathematical model. For example, consider

EXAMPLE 5. A fair coin is tossed twice. Let A and B be the events:

$$A = \text{"heads on first toss"}$$

$$B = \text{"heads on second toss."}$$

Often, we simply assert the independence of A and B based on our assumption that the result on the first toss can have no effect on the outcome of the second toss and vice versa. The reasoning here is useful (and, fortunately correct). In order to prove independence of A and B, we use the sample space

$$S = \{(H, H), (H, T), (T, H), (T, T)\}$$

and equiprobable probability measure P. Then

$$A = \{(H, H), (H, T)\}$$

$$B = \{(H, H), (T, H)\}$$

and

$$A \cap B = \{(H, H)\}.$$

Hence,

$$P(A) = \frac{n(A)}{n(S)} = \frac{2}{4} = \frac{1}{2},$$

$$P(B) = \frac{n(B)}{n(S)} = \frac{2}{4} = \frac{1}{2},$$

and

$$P(A \cap B) = \frac{n(A \cap B)}{n(S)} = \frac{1}{4}.$$

So, $P(A \cap B) = P(A)P(B)$, and independence follows. (Notice that A and B are not mutually exclusive because $A \cap B \neq \varnothing$.)

In the case of general events A and B we have

THEOREM 3. Let A and B be subsets of a finite sample space S with probability measure P.

Then

$$P(A \cup B) = P(A) + P(B) - P(A \cap B).$$

PROOF. The proof follows easily from a bit of set-theoretic trickery. (Easy to verify with Venn diagrams.) We have

$$A \cup B = A \cup (B \setminus A).$$

Moreover, $A \cap (B \setminus A) = \varnothing$, so part 3 of theorem 2 gives

$$P(A \cup B) = P(A \cup (B \setminus A))$$
$$= P(A) + P(B \setminus A).$$

Also,

$$B = (B \cap A) \cup (B \setminus A)$$

and

$$(B \cap A) \cap (B \setminus A) = \varnothing.$$

Again, we have from part 3 of theorem 2,

$$P(B) = P((B \cap A) \cup (B \setminus A))$$
$$= P(A \cap B) + P(B \setminus A).$$

Thus,

$$P(B \setminus A) = P(B) - P(A \cap B).$$

Hence

$$P(A \cup B) = P(A) + P(B \setminus A)$$
$$= P(A) + P(B) - P(A \cap B)$$

and the theorem is proved.

Part 3 of theorem 2 now becomes a special case of theorem 3. And either of these gives the result promised earlier:

COROLLARY 1. Let $A \subset S$. Then

$$P(A) + P(S \setminus A) = 1.$$

PROOF. $S = A \cup (S \setminus A)$ and $A \cap (S \setminus A) = \varnothing$. Hence

$$P(S) = P(A \cup (S \setminus A))$$
$$= P(A) + P(S \setminus A).$$

But $P(S) = 1$. So

$$1 = P(A) + P(S \setminus A).$$

CONDITIONAL PROBABILITY

A cube with dots on each of its six faces is called a die. Each face has either one, two, . . . , five, or six dots and they represent the numbers 1, 2, . . . , 6. The plural of die is dice. The gambling game called craps concerns itself with the sum of the numbers that appear face up when two dice are thrown. We will use phrases "outcome" or "the number that comes up" to refer to the face-up number on the first and second die or to the sum of the two numbers, depending on the context.

EXAMPLE 6: Two fair dice are thrown. Let A and B be the events:

A = "seven comes up" and
B = "the first die comes up 2 or 4."

Find $P(A)$ and $P(B)$.

SOLUTION. A typical outcome can be denoted by a pair of integers (n, m) when each of n and m is one of the numbers 1, 2, 3, 4, 5, 6. Then, for example, (2, 4) represents the outcome of "the first die comes up 2 and the second die comes up 4." The sample space S consists of the set of all such pairs:

$$S = \{(1, 1), (1, 2), (1, 3), (1, 4), (1, 5), (1, 6), (2, 1), \ldots, (6, 5), (6, 6)\}.$$

S contains exactly six pairs whose first element is 1, six pairs whose first element is 2, and so on. Thus $n(S) = 36$. The event A consists of all the pairs of S whose sum is 7. B contains those pairs whose first element is 2 or 4. So,

$$A = \{(1, 6), (2, 5), (3, 4), (4, 3), (5, 2), (6, 1)\}$$

and

$$B = \{(2, 1), (2, 2), \ldots, (2, 6), (4, 1), (4, 2), \ldots, (4, 6)\}.$$

Thus $n(A) = 6$ and $n(B) = 12$.

We use equiprobable measure because the dice are fair

$$P(A) = \frac{n(A)}{n(S)} = \frac{6}{36} = \frac{1}{6}$$

and

$$P(B) = \frac{n(B)}{n(S)} = \frac{12}{36} = \frac{1}{3}.$$

We have determined, in particular, that the probability of throwing a 7 with fair dice is $\frac{1}{6}$. Patrons of Las Vegas or Atlantic City will find this useful. But let's now consider a different question. Suppose two fair dice are tossed and we are given the information: "the first die comes up 2 or 4." Now, what is the probability that the two dice come up seven?

Notice that we are given information about the first die but none about the second die. More precisely, we can phrase our question as:

What is the probability the dice come up seven given that the first die comes up 2 or 4?

In terms of the events A and B of example 6 this becomes:

What is $P(A \mid B)$?

A method of answering the question follows:

> Since we are given the information that the initial die comes up 2 or 4, many of the outcomes belonging to the sample space S of example 5 are eliminated. The constrained sample space becomes

$$T = \{(2, 1), (2, 2), \ldots, (2, 6), (4, 1), \ldots, (4, 6)\}.$$

However, the dice remain fair so that each of the elements of T still occurs with equal probabilities. The subset of this that yields a total of seven is:

$$C = \{(2, 5), (4, 3)\}.$$

If we let Q denote the equiprobable probability measure on T we have

$$Q(C) = \frac{n(C)}{n(T)} = \frac{2}{12} = \frac{1}{6}.$$

This probability is "the probability of A given B." Thus

$$P(A \mid B) = \frac{1}{6}.$$

Now look at the event $A \cap B$ as a subset of S. We have

$$A \cap B = \{(2, 5), (4, 3)\}$$

and

$$P(A \cap B) = \frac{n(A \cap B)}{n(S)} = \frac{2}{36} = \frac{1}{18}.$$

Observe that

$$P(A \mid B) = \frac{1}{6} = \frac{\frac{1}{18}}{\frac{1}{3}} = \frac{P(A \cap B)}{P(B)}.$$

Thus, in this particular example,

$$P(A \mid B) = \frac{P(A \cap B)}{P(B)}.$$

It can be shown rather easily that the above formula for $P(A \mid B)$ holds in any finite sample space provided whenever we use this constrained sample space interpretation of "probability of A given B." Accordingly, the above formula is taken as the general definition of this concept.

DEFINITION 5. Let A and B be subsets of a finite sample space S with probability measure P. Suppose $P(B) \neq 0$. We define $P(A \mid B)$ by

$$P(A \mid B) = \frac{P(A \cap B)}{P(B)}.$$

and call it the conditional probability of A given B.

We see immediately that if A and B are independent events (see definition 4) then

$$P(A \mid B) = \frac{P(A) \cdot P(B)}{P(B)}.$$

Thus,

$$A \text{ and } B \text{ independent} \Rightarrow P(A \mid B) = P(A).$$

So, if A and B are independent then—in terms of conditional probability—A is unaffected by the occurrence of B.

Definition 5 immediately gives the equivalent and extremely useful equation

(4) $$P(A \cap B) = P(B) \cdot P(A \mid B).$$

This equation says: "the probability of A and B equals the probability of B multiplied by the conditional probability of A given B." When coupled with an appropriate tree diagram, this equation has vast applicability.

EXAMPLE 7. A box contains 3 black balls and 5 white balls. A ball is selected at random from the box and not returned. Then a second ball is selected from the box. What is the probability that the second ball is black?

Solution. The probability of getting a black ball on the first selection is $\frac{3}{8}$ since 3 of the 8 balls are black. If the ball was returned to the box before the next selection, then the probability of a black ball on the second trial would also be $\frac{3}{8}$. But we are using the process mathematicians refer to as *sampling without replacement*. The drawn ball is not returned. We cannot *assume* that the probabilities do not vary from trial to trial.

Figure 52 provides an illustration and shows a tree diagram to which "branch weights" have been appended.

$$P \text{ ("2nd ball is black")} = \frac{3}{8} \cdot \frac{2}{7} + \frac{5}{8} \cdot \frac{3}{7}$$

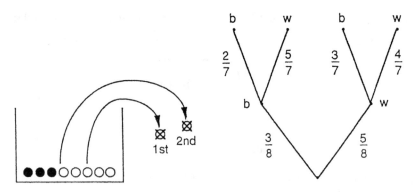

Figure 52: Sampling twice without replacement

The sample space consists equivalently of all paths through the tree or the set

$$S = \{(b, b), (b, w), (w, b), (w, w)\}.$$

(The symbol (b, w), for example, means: "1st ball selected is black and second is white.")

Let $A =$ "2nd ball is black." Then

$$A = \{(b, b), (w, b)\}.$$

We want to find $P(A)$. But, since the sampling is without replacement, P is not an equiprobable measure. We proceed as follows:

$$A = \{(b, b)\} \cup \{(w, b)\}.$$

Thus

(5) $$P(A) = P(\{(b, b)\}) + P(\{(w, b)\})$$

because the events $\{(b, b)\}$ and $\{(w, b)\}$ are mutually exclusive. Equation (4) gives:

$$P(\{(b, b)\}) = P(\text{"1st ball black and 2nd ball black"})$$
$$= P(\text{"1st ball black"}) \cdot P(\text{"2nd ball black}$$
$$\text{given first ball black"}).$$

But,

(6) $$P(\text{"1st ball black"}) = \frac{3}{8}.$$

since 3 of the original 8 balls are black. Moreover, if the first drawn ball is black, there remain 2 black balls and 5 white balls. Thus

$$P(\text{"2nd ball black given 1st ball black"}) = \frac{2}{7}$$

since two of the seven possibilities are black. Therefore,

$$P(\{(b,b)\}) = \frac{3}{8} \cdot \frac{2}{7}.$$

In figure 52 the branch weights $\frac{3}{8}$ and $\frac{2}{7}$ represent respectively, $P(\text{"1st ball black"})$ and $P(\text{"2nd ball black given 1st ball black"})$. The product of these two numbers equals the probability of the left-most path through the tree. This probability is $P(\{(b, b)\})$.

In exactly the same manner we see that

$$P(\{(w,b)\}) = \frac{5}{8} \cdot \frac{3}{7}.$$

Then by equation 5

$$P(A) = \frac{3}{8} \cdot \frac{2}{7} + \frac{5}{8} \cdot \frac{3}{7}$$

$$= \frac{6}{56} + \frac{15}{56}$$

$$= \frac{21}{56}$$

$$= \frac{3}{8}.$$

Thus

(7) $P(\text{"2nd ball black"}) = \frac{3}{8}.$

This solution method can be accomplished quickly and mechanically. Just draw a true diagram, attach branch weights, and mark the paths that give the desired outcome. The probability of each path equals the product of the branch weights. The probability of the desired outcome equals the sum of the marked paths.

Moreover, the process of solution and equation (4) generalize naturally to situations where several samples are drawn without replacement.

EXAMPLE 8. A box contains 3 black balls and 5 white balls. Three balls are randomly selected without replacement. What is the probability that the third ball is black?

SOLUTION. We need only extend the tree diagram of example 6 an additional stage. Figure 53 shows the new diagram with appended branch weights. The four paths marked with arrows constitute the desired outcome

$$B = \text{"3rd ball is black."}$$

So,

$$P(B) = \frac{3}{8} \cdot \frac{2}{7} \cdot \frac{1}{6} + \frac{3}{8} \cdot \frac{5}{7} \cdot \frac{2}{6} + \frac{5}{8} \cdot \frac{3}{7} \cdot \frac{2}{6} + \frac{5}{8} \cdot \frac{4}{7} \cdot \frac{3}{6}$$

$$= \frac{1}{56} + \frac{5}{56} + \frac{5}{56} + \frac{10}{56}$$

$$= \frac{21}{56} = \frac{3}{8}.$$

$$P\cdot(\text{"3rd ball is black"}) = \frac{3}{8}\cdot\frac{5}{7}\cdot\frac{2}{6} + \frac{3}{8}\cdot\frac{2}{7}\cdot\frac{1}{6} + \frac{5}{8}\cdot\frac{3}{7}\cdot\frac{2}{6} + \frac{5}{8}\cdot\frac{4}{7}\cdot\frac{3}{6}$$

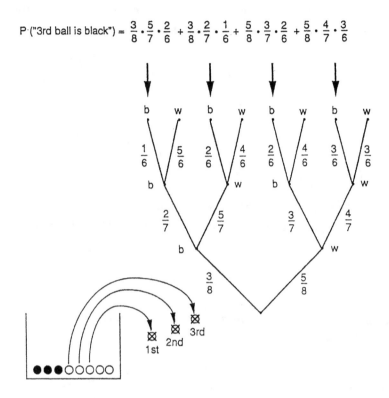

Figure 53: Sampling three times without replacement

Thus

(8) $$P(\text{"3rd ball black"}) = \frac{3}{8}.$$

Equations (6), (7), and (8) provide the unexpected result that the probability of drawing a black ball at any stage is the same and, of course, equals the probability of selecting a black ball on the first trial. Had we selected the three balls *with replacement*; that is, returned each drawn ball to the box after noting its color, then the probabilities of course would be the same. (At each trial the box would contain 3 black and 5 red balls. So the probability of a black ball is $\frac{3}{8}$.) However, we selected the three balls without replacement. Yet the probabilities remain unchanged. Is this a remarkable coincidence or is there something deeper going on? The answer is the latter.

THEOREM 3. A box contains n black balls and m white balls. Balls are randomly selected one by one, without replacement, until the box is empty. Let A_k denote the event:

$$A_k = \text{``the } k\text{th ball is black''}$$

for $k = 1, 2, \ldots, n + m$. Then

$$P(A_k) = \frac{n}{n + m}$$

for $k = 1, 2, \ldots, n + m$.

Partial proof. We will deal only with the case $k = 2$. Moreover—in order to avoid elaborate notation—we will use the tree diagram method for $k = 2$.

Clearly,

$$P(A_1) = \frac{n}{n + m}$$

since initially the box contains $n + m$ balls of which n are black.

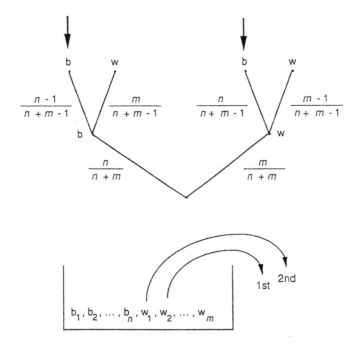

Figure 54: General case of sampling twice without replacement

Figure 54 shows the tree diagram at the second selection stage. The arrows mark the two paths that give a black ball on the second trial. $P(A_2)$ then equals the sum of the probabilities of these two paths. And the probability of each path is just the product of its branch weights. (The conditional probability appended to the second branch of the left-most path, for example, comes from the observation that if a black ball has been chosen at the first stage, then the box contains $n + m - 1$ balls, of which $n - 1$ are black.)

We have

$$P(A_2) = \left(\frac{n}{n+m}\right)\left(\frac{n-1}{n+m-1}\right) + \left(\frac{m}{n+m}\right)\left(\frac{n}{n+m-1}\right)$$

$$= \frac{n(n-1) + mn}{(n+m)(n+m-1)}$$

$$= \frac{n(n-1+m)}{(n+m)(n+m-1)}$$

$$= \frac{n}{n+m}.$$

This proves the theorem for $k = 2$.

The proof for the general case $k = 1, 2, \ldots, n + m$ is tedious but theoretically no more difficult. Notice that no use is made of the concept of "box" or "balls" in the proof. The result holds for any type of random nonreplacement sampling. The "box" might be a deck of cards and the "balls" individual cards.

EXAMPLE 8. Cards are dealt one at a time from an ordinary, well-shuffled deck of playing cards. Clearly,

$$P(\text{"1st card is the king of hearts"}) = \frac{1}{52}.$$

But theorem 3 ensures:

$$P(\text{"2nd card is the king of hearts"}) = \frac{1}{52}.$$

$$P(\text{"3rd card is the king of hearts"}) = \frac{1}{52}.$$

$$P(\text{``52nd card is the king of hearts''}) = \frac{1}{52}.$$

Interesting, don't you think?

COUNTING TECHNIQUES

In order to pursue probability theory you need on hand a variety of counting techniques. We will look at only two: one a method for counting *permutations* and the other a method for *combinations*. (Definitions of these terms follow.) The only tool we require is a result known as the *fundamental counting principle*.

From my house there are three running routes that lead to a place in the mountains called Wind Gap. From Wind Gap there are two running paths leading to Hawk Mountain. So, if I want to run to Hawk Mountain by way of Wind Gap there are $3 \cdot 2 = 6$ possible routes: 3 to Wind Gap and 2 to Hawk Mountain for each one of these. This simple idea generalizes to *the fundamental counting principle: If an event can occur in any one of* n *ways and then a second event can occur in any one of* m *ways, then the two events can occur in* n · m *ways.*

The principle itself generalizes in an obvious way to more than two events. For example, suppose you want to arrange 6 books in a row on a shelf. In how many ways can this be done?

We just use the counting principle repeatedly. The first position (on the left, say) can be filled with any one of the 6 books. Once this has been done, the second position can be filled with any of the remaining 5 books. Thus there are $6 \cdot 5 = 30$ ways to fill the first two slots. There remain 4 choices for the third position, and so on, down to a single remaining book for the last position. So the total number of ways to arrange the six books on the shelf is

$$6 \cdot 5 \cdot 4 \cdot 3 \cdot 2 \cdot 1 = 720.$$

The number on the left in the above equation is called "six factorial" and is written

$$6! = 6 \cdot 5 \cdot 4 \cdot 3 \cdot 2 \cdot 1.$$

In general, if *n* is any natural number then "*n* factorial" is defined by:

$$n! = n(n-1)(n-2)\ldots 3 \cdot 2 \cdot 1.$$

Also, by definition,

$$0! = 1.$$

An arrangement of n distinct objects in a line is called a *permutation* of the n objects. Arguing exactly as we did for the arrangement of books on a shelf, we obtain:

THEOREM 4. The number of permutations of n objects is $n!$.

In baseball, for example, a "batting order" is a list of the nine starting players, giving the sequence in which they come to bat. Thus a batting order is simply a permutation of the nine players. Theorem 4 tells us there are

$$9! = 9 \cdot 8 \cdot 7 \cdot 6 \cdot 5 \cdot 4 \cdot 3 \cdot 2 \cdot 1$$

possible batting orders.

If, however, we consider only the first three positions in the order, the counting principle ensures that there are $9 \cdot 8 \cdot 7 = 504$ possible choices here as the *number of permutations of 9 objects taken 3 at a time*. A standard symbol for this number is $P(9, 3)$. So,

$$P(9, 3) = 9 \cdot 8 \cdot 7.$$

In general, if we are given n distinct objects and a natural number k where $k \leq n$ we write

$$P(n, r) = \text{"number of permutations of n}$$
$$\text{objects taken r at a time."}$$

(Notice that the "P" in this symbol has nothing to do with probability.)

THEOREM 5.

$$P(n, r) = n(n - 1) \ldots (n - r + 1).$$

PROOF. $P(n, r)$ represents the number of ways we can arrange n distinct objects in a row of r slots. There are n choices for the first slot. Once this choice is made there are $n - 1$ choices for the second slot. Continuing in this manner gives:

$$n \text{ choices for the 1st slot}$$

$$n - 1 \text{ choices for the 2nd slot}$$

$$n - 2 \text{ choices for the 3rd slot}$$

. . .

$n - r + 1$ choices for the rth slot.

The counting principle ensures that the totality of choices equals the product of these. Thus,

$$P(n, r) = n(n - 1) \ldots (n - r + 1)$$

and the theorem is proved.

Notice that $r = n$ gives

$$P(n, n) = n(n - 1) \ldots 1 = n!$$

so that theorem 4 is a special case of theorem 5. Each theorem involves the concept of *lists* or *arrangements* or *order*. The symbol $P(n, r)$ stands for the number of ways n objects may be arranged in order when they are selected r at a time. When $r = n$ we get the number of ways this can be done when all of the objects are listed; that is, we get the number of permutations of the n objects.

When we deal with sets and subsets, however, the concept of order plays no role. The set $\{a, b, c\}$ and the set $\{c, b, a\}$ are identical because they contain exactly the same elements. But the *permutation abc* differs from the *permutation cba* precisely because permutations are defined in terms of order.

To make clear these matters, consider the set

$$S = \{a, b, c\}.$$

There are exactly 3 subsets of S that contain exactly 2 elements. They are:

$$\{a, b\}, \{a, c\}, \text{ and } \{b, c\}.$$

On the other hand, theorem 5 says there exist exactly

$$P(3, 2) = 3 \cdot 2 = 6$$

permutations of the three objects a, b, c when they are taken two at a time. These permutations are:

$$ab, ba, ac, ca, bc, \text{ and } cb.$$

Let's write

$$C(3, 2) = \text{``\textit{number of subsets of size 2 that can}}$$
$$\textit{be chosen from a set of size 3.''}$$

We know that the three-element set $S = \{a, b, c\}$ has exactly 3 subsets of size 2. Since the *number* of such subsets is independent of the names of the elements, *any set of size* 3 *has exactly* 3 *subsets of size* 2. Thus

$$C(3, 2) = 3.$$

Look at the relationship between $C(3, 2) = 3$ and $P(3, 2) = 6$. Namely,

$$P(3, 2) = 2 \cdot C(3, 2)$$

or, better still,

$$P(3, 2) = 2!C(3, 2).$$

This relationship shows clearly why $P(3, 2)$ exceeds $C(3, 2)$: *each subset of size* 2 *contains* 2 *elements that can be permuted in* 2! *ways.* For example, the subset $\{a, b\}$ of S has 2 elements a and b. And there are $2! = 2$ possible permutations: ab and ba.

These observations can be generalized. Let $C(n, r)$ be defined by

$$C(n, r) = \text{``number of subsets of size } r \text{ that can}$$
$$\text{be chosen from a set of size } n\text{''}$$

where $1 \leq r \leq n$.

THEOREM 6.

$$C(n, r) = \frac{n(n-1)(n-2) \ldots (n-r+1)}{r!}.$$

PROOF. The r objects in each subset of size r can be permuted in $r!$ ways by theorem 4. The number of such subsets is, by definition, $C(n, r)$. The fundamental counting principle tells us that the total number of permutations of the n objects in the given set taken r at a time will equal the product of the number of ways the objects can be selected and the number of ways that, once selected, they can be permuted. Therefore,

$$P(n, r) = C(n, r) \cdot r!.$$

Hence,

$$C(n, r) = \frac{P(n, r)}{r!}.$$

But theorem 5 gives

$$P(n, r) = n(n - 1)(n - 2) \ldots (n - r + 1).$$

Hence

$$C(n, r) = \frac{n(n-1)(n-2) \ldots (n-r+1)}{r!}$$

and the theorem is proved.

Probabilists think of the "C" in $C(n, r)$ as standing for "combination" and read $C(n, r)$ as "the number of combinations of n objects taken r at a time." However, I much prefer the "number of subsets" interpretation since this point of view makes clear the fact that the order of the objects plays no role.

Moreover, the numbers $C(n, r)$ can be written in the altered form

$$C(n, r) = \frac{n(n-1) \ldots (n-r+1)}{r!} \cdot \frac{(n-r)!}{(n-r)!}$$

$$= \frac{n(n-1) \ldots (n-r+1)(n-r)(n-r-1) \ldots 2 \cdot 1}{r!(n-r)!}.$$

Therefore,

$$C(n, r) = \frac{n!}{r!(n-r)!}.$$

In this form the numbers $C(n, r)$ are called "binomial coefficients" and they appear ubiquitously throughout mathematical analysis. In analysis they are denoted by the symbol $\binom{n}{r}$. That is

$$\binom{n}{r} = \frac{n!}{r!(n-r)!}.$$

These numbers take their name from the famous *binomial theorem*, which asserts that for any pair of real numbers and natural number n,

(9)
$$(x+y)^n = \sum_{r=0}^{n} \binom{n}{r} x^r y^{n-r}.$$

(One way to prove the binomial theorem is to fix $x, y \in \mathbb{R}$ and to let S denote the set of natural numbers for which (9) holds. Then show $S = \mathbb{N}$ by induction. The same method establishes the binomial theorem for complex numbers.)

In the special case $n = 3$, equation (9) gives

(10) $\quad (x+y)^3 = \sum_{r=0}^{3} \binom{3}{r} x^{3-r} y^r = \binom{3}{0} x^3 y^0 + \binom{3}{1} x^2 y^1 + \binom{3}{2} x^1 y^2 + \binom{3}{3} x^0 y^3.$

We now have before us a plethora of ways in which to interpret the numbers $\binom{3}{r}$:

(i) $\quad \binom{3}{r} =$ "number of subsets of size r that can be chosen from a set of size 3"

(ii) $\quad \binom{3}{r} = \dfrac{3!}{r!(3-r)!}$

(iii) $\quad \binom{3}{r} =$ "coefficient of $x^r y^{3-r}$ in the expansion of $(x + y)^3$

Using the first interpretation, we know already that

$$\binom{3}{2} = C(3,\ 2) = 3.$$

But

$$\binom{3}{1} = \binom{3}{2},$$

since each time we select from $\{a, b, c\}$ a subset of size 2 we leave behind exactly one subset of size 1, and conversely. So,

$$\binom{3}{1} = 3.$$

Clearly,

$$\binom{3}{0} = 1 \text{ and } \binom{3}{3} = 1.$$

($\{a, b, c\}$ has only one subset, the empty set, of size 0, and only one subset, itself, of size 3.)

As for interpretation (iii), we have

$$
\begin{aligned}
(x+y)^3 &= (x+y)(x+y)^2 \\
&= (x+y)(x^2 + 2xy + y^2) \\
&= x^3 + 2x^2y + xy^2 + yx^2 + 2xy^2 + y^3 \\
&= x^3 + 3x^2y + 3xy^2 + y^3.
\end{aligned}
$$

If we compare this expansion with equation (10) we can read off the values:

$$
\binom{3}{0} = 1, \ \binom{3}{1} = 3, \ \binom{3}{2} = 3, \text{ and } \binom{3}{3} = 1.
$$

These numbers can, of course, also be determined directly from (ii). Moreover, this complete discussion can be extended to arbitrary binomial coefficients $\binom{n}{r}$. What, for example, is the number $\binom{20}{3}$? Depends on your point of view:

(a) It is the number of subsets of size 3 you can select from a set of size 20.

(b) It is the number

$$
\frac{20!}{3!\,17!} = \frac{20 \cdot 19 \cdot 18}{3!} = 20 \cdot 19 \cdot 3 = 1{,}140.
$$

(c) It is the coefficient of x^3y^{17} in the expansion of $(x+y)^{20}$.

Or, from a practical point of view, if you think of a faculty as a set and a committee as a subset, then:

$$
\binom{20}{3} = \text{``number of three-person committees that can be selected from a mathematics department having 20 faculty members.''}
$$

But this interpretation scares me because the number is so large. No way I can escape 1,140 committees.

From the expression

$$\binom{n}{r} = \frac{n!}{r!(n-r!)}$$

we see

$$\binom{n}{0} = \frac{n!}{0!(n-0)!}$$

$$= \frac{n!}{n!}$$

$$= 1.$$

Similarly,

$$\binom{n}{n} = 1.$$

(Of course, $\binom{n}{0} = 1$ and $\binom{n}{n} = 1$ follow from the subset interpretation: a set of size n has exactly one subset with no elements and one subset with n elements.)

A straightforward calculation (which we will omit) shows

(11)
$$\binom{n}{r} + \binom{n}{r-1} = \binom{n+1}{r}.$$

(An easy way to verify equation (10) is to "evaluate" each binomial coefficient by using

$$\binom{p}{q} = \frac{p!}{q!(p-q)!}$$

and then showing both sides of (11) are identical.)

Figure 55 shows the binomial coefficients written in an array called *Pascal's triangle*. The general form of the triangle appears on the left of figure 55. The triangle on the right shows specific values. The left boundary of the triangle consists of all 1's because $\binom{n}{0} = 1$. Since $\binom{n}{n} = 1$, the right boundary also consists of all 1's. The other entries are then determined (in an obvious manner) through the use of equation (11).

$$\begin{array}{ccccccccccc}
& & & & & \binom{0}{0} & & & & & \\
& & & & \binom{1}{0} & & \binom{1}{1} & & & & \\
& & & \binom{2}{0} & & \binom{2}{1} & & \binom{2}{2} & & & \\
& & \binom{3}{0} & & \binom{3}{1} & & \binom{3}{2} & & \binom{3}{3} & & \\
& \binom{4}{0} & & \binom{4}{1} & & \binom{4}{2} & & \binom{4}{3} & & \binom{4}{4} & \\
\binom{5}{0} & & \binom{5}{1} & & \binom{5}{2} & & \binom{5}{3} & & \binom{5}{4} & & \binom{5}{5}
\end{array}$$

$$\begin{array}{ccccccccccc}
& & & & & 1 & & & & & \\
& & & & 1 & & 1 & & & & \\
& & & 1 & & 2 & & 1 & & & \\
& & 1 & & 3 & & 3 & & 1 & & \\
& 1 & & 4 & & 6 & & 4 & & 1 & \\
1 & & 5 & & 10 & & 10 & & 5 & & 1
\end{array}$$

Figure 55: Pascal's triangle

Binomial coefficients interest mathematical analysts, namely, because of their importance in the theory of functions. But even analysts commonly read the symbol $\binom{n}{r}$ as "n choose r" as a reminder of its counting role in set theory. And one type of set that $\binom{n}{r}$ will count is a set of cards called a poker hand.

EXAMPLE 10: Five cards are dealt fairly from an ordinary deck of playing cards.

(a) What is the probability that the five cards are all hearts?

(b) What is the probability that the five cards contain four aces?

SOLUTION: (a) Since we are not interested in the order in which the cards are dealt, we think of the five cards—the poker hand—as a subset of the complete deck. The sample space S then consists of the collection of all subsets of size 5 that can be selected from the complete deck, which is a set of size 52. We know there are $\binom{52}{5}$ such subsets. Thus

$$n(S) = \binom{52}{5}.$$

Since the cards are fairly dealt, the equiprobable probability measure P applies. So if A is the event

$$A = \text{"the hand contains 5 hearts"}$$

then

$$P(A) = \frac{n(A)}{n(S)}$$

$$= \frac{n(A)}{\binom{52}{5}}.$$

We need only determine $n(A)$. But $n(A)$ is the number of ways we can select 5 hearts from the total number of hearts in the deck. Since the deck contains 13 hearts, we have

$$n(A) = \binom{13}{5}.$$

Then,

$$P(A) = \frac{\binom{13}{5}}{\binom{52}{5}}.$$

It turns out that

$$\binom{52}{5} = 2,598,960$$

and

$$\binom{13}{5} = 1,287.$$

Hence,

$$P(A) = \frac{1,287}{2,598,960} = .0004952.$$

Card players call the hand of event A a "flush"—more particularly, a heart flush. We've shown

$$P(\text{"heart flush"}) = .0004952$$

which is quite small. You may wait a long time before the dealer gives you one.

(b) The sample space and the equiprobable measure remain. Let B be the event

$$B = (\text{"the hand contains four aces"}).$$

The four aces in the five-card hand can be chosen in $\binom{4}{4}$ ways. Then the remaining card can be chosen in $\binom{48}{1}$ ways. (Choose any card from the 48 non-aces in the deck.) The fundamental counting principle then gives:

$$n(B) = \binom{4}{4}\binom{48}{1}.$$

Hence,

$$P(B) = \frac{\binom{4}{4}\binom{48}{1}}{\binom{52}{5}}$$

$$= \frac{1 \cdot 48}{2,598,960}$$

$$= .0000185.$$

You'll wait even longer for this hand.

Many probability questions can be interpreted in the context of what are called *repeated Bernoulli trials*. A Bernoulli trial is simply an experiment with two possible outcomes, S and F (usually thought of as "success" and "failure"). Outcome S occurs with probability p and F with probability

$1 - p$. Thus we know the probability of obtaining outcome S on a single trial. We want to determine the probability of obtaining a given number of S's when the experiment is independently repeated.

THEOREM 7. A Bernoulli trial has two possible outcomes, S and F. S occurs with probability p and F with probability $1 - p$. The trial is independently repeated n times. Then the probability of exactly k successes, where $k = 0, 1, 2, \ldots, n$, is

$$\binom{n}{k} p^k (1 - p)^{n-k}.$$

PROOF: A particular outcome of the n repetitions of the trial may be represented by the symbol $(x_1, x_2, x_3, \ldots, x_n)$ where each x_j is S or F.

The sample space for the n repetitions then is

$$S = \{(x_1, x_2, \ldots, x_n) : x_j \in (\{S, F\}, j = 1, 2, \ldots, n\}.$$

For fixed $k = 0, 1, 2, \ldots, n$ let

$$A_k = \{(x_1, x_2, \ldots, x_n) : \text{exactly k of the } x_j\text{'s are } S\}.$$

Then A_k is the event

$$A_k = \text{"exactly } k \text{ } S\text{'s occur in the } n \text{ trial repetitions."}$$

We want to determine $n(A_k)$.

It is convenient to think of a typical member of A_k as a path through the tree diagram corresponding to the n repetitions of the Bernoulli trial. Figure 56 shows a typical path. The branch leading to each S has weight p, branches leading to F's have weight $1 - p$. (Since the trials are independent of one another, the remark following definition 5 tells us that the conditional probabilities that are ordinarily appended to the branches reduce to the actual probabilities. Here, they are either p or $1 - p$.)

The probability of the path shown in figure 56 is the product of the branch weights. But this product is

$$p^k (1 - p)^{n-k}$$

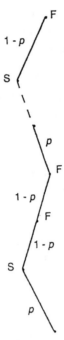

Figure 56: Path through tree diagram having k S's and $n - k$ F's

since the path contains k S's and, hence $(n - k)$ F's. But this path is a typical element of A_k. Then each element of A_k occurs with the same probability, $p^k(1 - p)^{n - k}$. So

$$P(A_k) = n(A_k) \cdot p^k(1 - p)^{n - k}.$$

We now must determine $n(A_k)$.

But each member of A_k is determined by the choice of positions in the symbol $(x_1, x_2, \ldots, x_n)$ when the x's become S. And there are exactly k such positions. We can choose these positions in exactly $\binom{n}{k}$ ways. Hence

$$n(A_k) = \binom{n}{k}.$$

Therefore,

$$P(A_k) = \binom{n}{k} p^k (1 - p)^{n - k}.$$

This proves the theorem.

The appellation "Bernoulli trial" is nothing more than a fancy name for a single toss of a coin—just think of heads as "success" and tails as "failure." Then theorem 7 says:

> *If a coin, which lands heads with probability* p, *is tossed* n *times independently, then the probability of exactly* k *heads equals*
>
> $$\binom{n}{k} p^k (1-p)^{n-k}.$$

EXAMPLE 11. A fair coin is tossed 20 times. What is the probability of exactly 10 heads?

SOLUTION. We assume the tosses are independent of one another and use theorem 7 with $p = \dfrac{1}{2}$, $n = 20$, and $k = 10$. Thus the probability of exactly 10 heads is

$$\binom{20}{10}\left(\frac{1}{2}\right)^{10}\left(1 - \frac{1}{2}\right)^{20-10} = \binom{20}{10}\left(\frac{1}{2}\right)^{20}.$$

The answer to example 11 turns out to be

$$\binom{20}{10}\left(\frac{1}{2}\right)^{20} = \frac{20!}{10!10!}\left(\frac{1}{2}\right)^{20}$$

$$= \frac{20 \cdot 19 \cdot 18 \ldots 11}{10!}\left(\frac{1}{2}\right)^{20}$$

$$= .1762.$$

The probability is unexpectedly small. If someone regularly obtains exactly 10 heads in each set of 20 tosses of his "fair coin," then it probably isn't.

EXPECTATIONS

A real-valued function defined on a sample space is called a *random variable*. So if S is a sample space and

$$f: S \to \mathbb{R}$$

then f is a random variable. Much of probability can be stated in terms of the behavior of certain random variables. Consider, for example, the Bernoulli trial situation described in theorem 7.

EXAMPLE 12. A Bernoulli trial has possible outcomes S and F, which occur with probabilities p and $1 - p$, respectively. The trial is repeated independently n times. For each outcome $x = (x_1, x_2, \ldots, x_k)$ in the sample space S, let

f = "number of S's in $(x_1, x_2, x_1, x_2, \ldots, x_k)$."

Thus f is a random variable that counts the numbers of success in the n repetitions of the Bernoulli trial. The range of f is

$$R(f) = \{0, 1, 2, 3, \ldots, n\}$$

and theorem 7 gives the probabilities with which these values are attained. We may write

$$P(f = k) = \binom{n}{k} p^k (1 - p)^{n-k}$$

for $k = 0, 1, 2, \ldots, n$, where the symbol $P(f = k)$ means: "probability that $f = k$" or "probability of exactly k S's."

In the special case of example 11 when $p = \dfrac{1}{2}$, $n = 20$, we have

$$R(f) = \{0, 1, 2, \ldots, 20\}.$$

The result of example 12 says

$$P(f = 10) = \binom{20}{10} \left(\frac{1}{2}\right)^{20}.$$

Random variables are particularly applicable to games of chance.

EXAMPLE 13. A roulette wheel has 20 slots into which a ball can roll. A player bets $1.00 on a particular numbered slot. If the ball rolls into the chosen slot. the player wins $35.00. Otherwise, the house takes the player's $1.00.

Let f be the random variable that counts the player's "winnings." The range of f is

$$R(f) = \{-1, 35\}$$

because, in dollars, his winnings are either 35 or −1. Since the wheel has 38 slots and only one of them gives the player a win, we have:

$$P(f = 35) = \frac{1}{38}$$

and

$$P(f = -1) = \frac{37}{38}.$$

In real-world situations, probabilities are often interpreted in terms of frequency. That is, if an event occurs with probability p, then we expect the number p to represent the fraction of times the event actually appears. (A theorem called the *law of large numbers* asserts the validity of this interpretation in the strict mathematical sense.)

So, here—about one time out of each 38 tries—the player can expect to win $35. He expects to lose about 37 times out of each 38 attempts. Therefore, if he plays roulette in this manner time after time, he expects to win on each try:

$$u = (-1)\left(\frac{37}{38}\right) + (35)\left(\frac{1}{38}\right)$$

$$= -\frac{2}{38}$$

$$= -\frac{1}{19}.$$

Thus the player expects to lose approximately one nickel each time he plays the game. With any luck, he'll go broke slowly.

The number u that appears in example 13 bears the name "the expected value of the random variable f" or "the expectation of f."

DEFINITION 6. Let f be a random variable on a finite sample space S so that

$$f: S \to \mathbb{R}.$$

Let $R(f)$ denote the range of f. We define the expectation of f or the expected value of f by

$$u = E(f) = \sum_{y \in R(f)} yP(f = y).$$

Thus the expected value of a random variable is the number obtained by multiplying each possible value of f by the corresponding probability with which it is attained and summing over all these values.

EXAMPLE 14. Toss a fair coin twice. Let

$$f = \text{"number of heads that come up."}$$

Then
$$R(f) = \{0, 1, 2\}$$

and (by theorem 7)

$$P(f = 0) = \binom{2}{0}\left(\frac{1}{2}\right)^2 = \frac{1}{4}$$

$$P(f = 1) = \binom{2}{1}\left(\frac{1}{2}\right)^2 = \frac{1}{2}$$

$$P(f = 2) = \binom{2}{2}\left(\frac{1}{2}\right)^2 = \frac{1}{4}.$$

Then
$$E(f) = \sum_{y \in R(f)} yP(f = y)$$

$$= \sum_{k=0}^{2} k\binom{2}{k}\left(\frac{1}{2}\right)^k$$

$$= 0 \cdot \frac{1}{4} + 1 \cdot \frac{1}{2} + 2 \cdot \frac{1}{4}$$

$$= \frac{1}{2} + \frac{1}{2}$$

$$= 1.$$

So, the "expected number of heads" in two tosses of a fair coin is 1.

Probabilists know things the rest of us do not. Often they can, by a combination of physical and probabilistic reasoning, arrive easily at mathematical expressions that can be verified only through complicated analysis. Here's an example.

EXAMPLE 15. A biased coin lands heads with probability p. If the coin is tossed n times, it is reasonable to "expect" about np heads—the number of tosses n multiplied by the fraction of times p that a head should appear.

Thus if f is the random variable that counts the heads,

$$f = \text{"number of heads that appear"}$$

then we should have—on purely intuitive grounds

$$E(f) = np.$$

On the other hand,

$$R(f) = \{0, 1, 2, \ldots, n\}$$

and

$$P(f = k) = \binom{n}{k} p^k (1-p)^{n-k}$$

by theorem 7. The definition of $E(f)$ then gives

$$E(f) = \sum_{k=0}^{n} k \binom{n}{k} p^k (1-p)^{n-k}$$

(the values of f multiplied by the corresponding probabilities and their summand). Therefore, our intuitive probabilistic argument suggests the identity

(12) $$\sum_{k=0}^{n} k \binom{n}{k} p^k (1-p)^{n-k} = np.$$

Equation (12) is, in fact, true. It can be proved by purely analytic methods that have nothing to do with probability theory. Any mathematician can prove equation (12). But a probabilist finds it obvious.

In passing, we should note that the expected value of a random variable may be attained only with low probability. If we toss a fair coin 20 times the equation (12) gives

$$E(f) = 20 \cdot \frac{1}{2} = 10.$$

But example 10 gave (in the new language) the relatively small probability

$$P(f = 10) = \binom{20}{10}\left(\frac{1}{2}\right)^{20} = .1762.$$

In fact, the expected value may never be attained at all. Toss a fair coin once. Then

$$E(f) = np = 1 \cdot \frac{1}{2} = \frac{1}{2}.$$

Tough to get half of a head with any coin, fair or not.

STATISTICS

I think of statistics as the inverse of probability theory. Statistical questions are essentially probability questions asked the other way around. A probabilist might ask the question:

What is the probability of obtaining 49 heads in 100 tosses of a fair coin?

A statistician, on the other hand, wants to know:

If 100 tosses of a coin yield 49 heads, can we infer that the coin is fair?

Thus the probabilist begins with knowledge of the probability p of heads on a single toss of the coin: $p = \frac{1}{2}$ because the coin is fair. The statistician, however, is given only data. He knows that 100 tosses produced 49 heads. Is it reasonable, the statistician wants to know, to draw the conclusion $p = \frac{1}{2}$?

To be sure, the true nature of statistics—and of probability theory—is more complicated than this simple example shows. But this illustrates the gist of the two areas. Moreover, real-world statistical problems are rarely phrased in terms of coin tosses. But they might be. Consider, for example, a preelection poll.

EXAMPLE 16. The 80 members of a certain political organization are to choose between two candidates for the position of representative to the national organization. Prior to the election an informal poll is taken.

A random selection of 20 senate members is made and each is asked: "Will you vote for Candidate *A*?" Exactly 10 of them answer: "Yes." What can you conclude?

SOLUTION. Obviously, the election will be close. The relatively large sample of one-fourth of the entire voting population turned up dead even.

Let's consider example 15 a bit more precisely. A certain fraction of the faculty will vote for Candidate *A* on election day. Let's call this fraction *p*. Of course, *p* is unknown until after the election passes and the votes are counted. The purpose of the polling process is to provide preelection information regarding the value of *p*.

Think of the process as the tossing of a biased coin. Our basic *assumption* is this: an arbitrary voter will—when either voting or being polled—or not choose Candidate *A* by tossing a coin that lands heads with probability *p*. If the coin lands heads, he selects *A*.

Our polling process then may be considered as the result of twenty tosses of this coin. The twenty tosses produced exactly ten heads, which leads us to conclude that $p = \dfrac{1}{2}$. To see the reasonable validity of this conclusion consider the following:

Let *f* be the random variable that gives the number of heads provided by twenty tosses of our coin. Thus the range of *f* is:

$$R(f) = \{0, 1, 2, \ldots, 20\}$$

and

$$P(f = k) = \binom{20}{k} p^k (1 - p)^{20-k}.$$

If $p = \dfrac{1}{2}$, that is, the coin is fair, then we have

$$P(f = k) = \binom{20}{k}\left(\frac{1}{2}\right)^k\left(1 - \frac{1}{2}\right)^{20-k}$$

$$= \binom{20}{k}\left(\frac{1}{2}\right)^{20}.$$

Some particular values turn out to be:

$$P(f = 8) = .120$$
$$P(f = 9) = .160$$
$$P(f = 10) = .176$$
$$P(f = 11) = .160$$
$$P(f = 12) = .120.$$

This list of numbers sums to .736. Thus

$$\sum_{k=8}^{12} P(f = k) = .736$$

which says

$$P(8 \leq f \leq 12) = .736.$$

On the other hand, suppose the true (but unknown) value of p is $p = \dfrac{1}{4}$, so that 75 percent of the voters will actually vote for Candidate B. In this case, the coin-tossing model gives:

$$P(f = k) = \binom{20}{k}\left(\frac{1}{4}\right)^k\left(1 - \frac{1}{4}\right)^{20-k}.$$

Here, the particular values become

$$P(f = 8) = .609$$
$$P(f = 9) = .0270$$
$$P(f = 10) = .0100$$
$$P(f = 11) = .0030$$
$$P(f = 12) = .0007.$$

This gives (to 3 decimal places)

$$\sum_{k=8}^{12} P(f = k) = .102$$

or

$$P(8 \le f \le 12) = .102.$$

In terms of the coin-tossing model, we see that the polling process of example 16 would produce an outcome within the range 8, 9, 10, 11, or 12, with a probability of .736 if the true value of p is $p = \frac{1}{2}$. But, if $p = \frac{1}{4}$, the polling outcome falls in the same range with the much smaller probability of .102.

The actual polling process gave a value of 10, which falls in the exact center of the given range. Thus it is clear that, between the two choices:

$$H_0 : p = \frac{1}{2}$$

and

$$H_1 : p = \frac{1}{4}$$

we choose the former. The coin-tossing model of the sample process leads us to *select* the hypothesis H_0 and to *reject* the hypothesis H_1.

Much of statistics concerns itself with the acceptance or rejection of a certain hypothesis. The process by which a particular hypothesis is considered often involves random sampling and the examination of the sample data in the light of probability considerations, more complicated than—but similar in nature to—the coin-tossing model. Notice that we looked at the probability of sampling outcomes falling within a given *range* rather than their taking on a *particular value*. This normally is the case since particular values are often unlikely. Example 10, in fact, points out the "smallness" of $P(f = 10)$ when $p = \frac{1}{2}$.

INFINITE SAMPLE SPACES

In all of our discussions we have considered only *finite* sample spaces. But infinite sample spaces arise naturally. Consider, for example, the experiment: toss a fair coin until it lands heads. A possible sample space for this experiment is:

$$S = \{H, TH, TTH, TTTH, TTTTH, TTTTTH, \ldots\}.$$

The symbol "*H*" stands for the outcome: "the first toss lands heads." "*TH*" means "first toss tails, second toss heads." "*TTH*" means "first toss tails, second toss tails, third toss heads." The other symbols are similarly defined. We stop tossing the coin when a heads appears. But the sample space is infinite because we must allow for the probability that the coin always comes up tails.

If we try to extend definition 2 to general countably infinite sample spaces such as

$$S = \{x_1, x_2, x_3, \ldots\}$$

we face an immediate complication. The condition of definition 2 that requires $w(x_1) + w(x_2) + \ldots\ w(x_n) = 1$ now becomes

(13) $$w(x_1) + w(x_2) + \ldots = 1.$$

The three dots in equation (13) indicate that the sum goes on forever. Such an infinite sum is called an *infinite series* and an understanding of these mathematical objects requires (as a minimum) a knowledge of calculus.

Uncountably infinite sample spaces also arise. Consider, for example, the experiment: select a random real number between 0 and 1. Here,

$$S = \{x : 0 \leq x \leq 1\}$$

and, as we have seen, S is not countable. Consequently, the members of S cannot be written in a list and even expressions like equation (13) are insufficient to describe the appropriate generalization of definition 2.

Such a generalization requires the calculus notion of an *integral*. In the probability theory of countably infinite spaces, sums become integrals and probabilities are interpreted as areas under certain curves known as *probability distributions*.

The best known of these curves belonga to a class called *normal distributions*. And, within this class, the most famous is the *standard normal dis-*

tribution. This bell-shaped curve is the graph of a function g whose domain of definition consists of the entire set of real numbers. The function is

$$g(x) = \frac{1}{\sqrt{2\pi}} e^{-\frac{x^2}{2}}.$$

Figure 57 shows the graph of this curve.

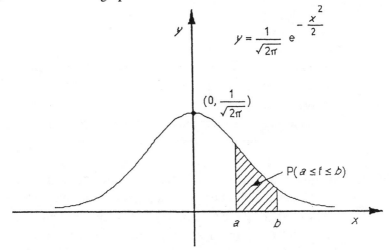

Figure 57: The standard normal distribution

It can be shown by calculus methods that the area under this curve is 1. If a random variable f has the standard normal distribution then it turns out that $P(a \leq f \leq b)$ becomes equal to the area under the curve between the vertical lines $x = a$ and $x = b$. This probability appears as the shaded area in figure 57.

Much of the importance of this distribution stems from the truly remarkable fact that the average of most collections of random variables becomes a new random variable whose distribution is approximately normal provided the collection is sufficiently large. The deep theorem that precisely presents this result is known as the *central limit theorem.* It is this theorem that allows statisticians to do their work. The central limit theorem ensures that the frequently made assumption that "data are normally distributed" is, at least, approximately correct provided the sample size is sufficiently large. Because of the central limit theorem, the standard normal distribution plays a major role in statistics and, hence, in applications of statistics. Consequently, because of the daily omnipresence of statistical applications, this distribution affects us all. And—because *e* and π appear in the equation of the standard normal distribution—these strange transcendental numbers affect us too.

Mathematicians tell this tale:

A certain man receives notice of an increase in his life insurance premiums.
He goes to his agent for explanation.

"You are older now," the agent says. "This places you in a more expensive category."
"Why?" the man asks.
"I'm not sure," the agent tells him. "Our actuaries figure this stuff out with mathematics. Ask them."

Determined, the man goes to the company headquarters and rides the elevator to the top floor, up to the penthouse where the actuaries work their magic. One of them tells him that insurance costs are based on life expectancy and that they determine expectancy with probability distributions, particularly the standard normal distribution.

"What the heck is a standard normal distribution?" the man asks.
"It's this," the actuary says, showing the man a graph of the thing.
"What are those symbols?"
"That's an equation that describes the curve. The standard normal distribution is the graph of this equation."
"OK but what is that?" asks the man, pointing.
"That's a number called *e*.
"And that?"
"The number π," says the actuary.
"But what are they?"
"*e* is hard to describe," the actuary says, "but π is the ratio of the circumference of a circle to its diameter."

"Would you repeat that?" the man asks patiently.

The actuary repeats the explanation.

"That's crazy," the man says, "what can circles have to
do with how long I will live?"

Crazy perhaps. But true nevertheless. And to understand it, you must understand mathematics.

Chapter 8

CALCULUS

The subject we call "calculus" arose from the combining of two apparently dissimilar notions: the concept of *area of a region bounded by curved lines* and the idea of *instantaneous velocity of a moving particle*. The first concept is the more ancient. Important contributions to its development were made by the early Greek mathematicians, particularly by Archimedes (287–212 BCE). Led mainly by Isaac Newton (1624–1727), the study of the idea of instantaneous velocity flourished in the seventeenth century. Only later, through the work of Augustin Cauchy (1789–1857) and Karl Weierstrass (1815–1897), were the notions made mathematically rigorous. When that happened, modern calculus popped into existence.

Many others participated in the development of the calculus and even a cursory outline of the history of the subject would fill a good-sized volume. But the men mentioned above played significant roles, and even a brief description of their work—together with that of Gottfried Leibniz (1646–1715)—provides an idea of the structure of this powerful and elegant branch of mathematics.

Let's look first at the problem of determining the area of a region bounded by curves. Figure 58 shows four planar regions bounded, respectively, by a rectangle, a right triangle, an "arbitrary" triangle, a circle, and an "arbitrary" closed curve. In the figure, the areas of the regions enclosed by the rectangle and by the triangle are determined by the well-known formulas "area equals length times width" and "area equals one-half the base times height." Initially, we need to consider the origin of these two formulas.

Before you can talk about "finding the area of a region," you must decide exactly what is meant by the notion of "area." This is easily accomplished in the case of the rectangle by taking the formula "area = length times width" to be the *definition* of area. This definition provides a mathematical formulation of our intuitive notion of what the area enclosed by a rectangle should be and agrees with measurement of a real-world area that can be obtained by, say, the laying of tiles on a rectangular floor. Let's suppose the rectangular area has been defined in this manner.

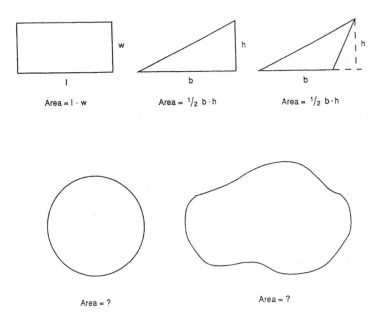

Figure 58: Areas

Once this has been done, then one can *prove* that the area of a right triangle is given by the formula in figure 58. This process could have, in fact, been done the other way around since the rectangle may be broken into the union of two right triangles. The triangle is, in this sense, the more *fundamental* of the two figures. However, as far as area is concerned, the rectangle is more *natural*.

The next step consists of proving that the formula "$A = \dfrac{1}{2} b \cdot h$" also holds for determining the area of an arbitrary triangle as shown in figure 58.

Let's suppose this has been done.

So far, everything has been straightforward. But now we must deal with the circle. If we put ourselves in the position of the ancient Greeks and look at the situation without benefit of our years of schooling, which has placed a formula in our heads—probably without associated understanding—we see that we do not yet have a *definition* of the notion of area enclosed by the circle. How can we *find* this area without first *defining* exactly what we are after? Part of the genius of the approach of Archimedes was that he instinctively did the two things simultaneously. This approach (described in the following) is called the *method of exhaustion.*

Look at figure 59. Here, the circle has been redrawn several times where, at each stage, a "polygon" has been inscribed. Each polygon is determined by dividing the circle into an even number of points. The number of sides of the polygon increases as the diagram advances. And in the "final" sketch, the polygon has been decomposed into triangles. Notice that, as the number of sides of the inscribed polygon increases, the polygon more and more closely approximates the circle. Consequently, the area enclosed by the inscribed polygon more and more nearly approximates the "area enclosed by the circle," whatever the meaning of this phrase might be. Moreover, the polygon may be decomposed into triangles in a natural manner. And, since the formula for the area of a triangle is known to us, we can compute the area enclosed by the polygon by "adding up" these smaller areas.

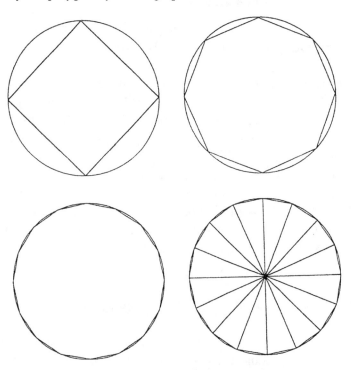

Figure 59: Method of exhaustion

Let us now suppose two things. First, suppose we have repeated figure 59 to produce a sequence of drawings where, from one step to the next, we inscribe polygons with an increasing number of sides. Let's denote these polygons by $P_1, P_2, P_3, \ldots, P_n$, where P_n, denotes the polygon produced at the nth stage of the repetition. At each step we can compute the area of the inscribed polygon. Let's call these resulting values $A_1, A_2, A_3, \ldots A_n$. Notice that our construction process causes each inscribed polygon to be contained in the next one. Thus, the areas of the polygons are getting larger as the number of sides increases. That is

$$A_1 < A_2 < A_3 < \ldots < A_n.$$

Second, suppose that we have *already* given some meaning to the notion of area enclosed by the circle. Let's say it has the value A. Then certainly for each inscribed polygon being contained in the circle, we have area less than the area of the circle. Consequently

(C) $A_1 < A_2 < A_3 < \ldots < A_n < A.$

Let's summarize. We want to compute something, namely, the area of a circle, which has not yet been defined. But, whatever it is, it must—on intuitive grounds—satisfy the set of inequalities given in (C). Moreover, (C) gives *increasing* sequences of numbers that never exceed our unknown quantity A. As n increases indefinitely and we consider more and more inscribed polygons, each one having twice as many sides as its predecessor, the numbers A_n do one of two things: they "approach" some fixed number more and more closely or else they do not.

Archimedes proved they, in fact, *do approach some fixed number.* This fixed number is *defined* to be the area of the circle. Hence, the number A, which we actually wrote in (C) before we were entitled to, is defined to be the *limit* of the sequence of numbers on the left-hand side as n increases without bound. In modem notation we write

$$A = \lim_{n \to \infty} A_n.$$

This equation is read: "A equals the limit as n approaches infinity of A sub $- n$."

The limit concept, expressed in this equation, lies at the core of calculus and represents the subject's central notion. We will have more to say about it later. For now, think of it intuitively as meaning that the variable A_n gets closer and closer to the *constant* A as n becomes arbitrarily large.

Archimedes proved, using this method, that there exists a constant that, when multiplied by the square of the radius of an arbitrary circle, yields its area. The constant, of course, is the famous number π. So, if A denotes the area of a circle of radius r, then $A = \pi r^2$.

The method can also be applied to the problem of determining, and simultaneously defining, the "length" of the circumference of an arbitrary circle. For a circle of radius r, the result is $C = 2\pi r$, where C denotes this length. Equivalently, $C = \pi d$, where d denotes the diameter of the given circle.

The symbol π, then, denotes a number that equals the ratio of the circumference of *any* circle divided by its diameter. This mathematical fact Archimedes established 2,200 years ago.

The number π turns out, as I have already mentioned, to be irrational (actually transcendental) and, hence, cannot be represented by a finite or repeating decimal. A four-decimal approximation to the value is 3.1416. Archimedes showed that it lives between the numbers 3.1408 and 3.1429.

All this Archimedes obtained from his method and its refinements. This brilliant and elegant technique represents, in fact, the beginnings of what became known, centuries later, as the *integral calculus*. But Archimedes proceeded intuitively and part of the accomplishments of Cauchy and Weierstrass in the nineteenth century was the establishment of rigorous meaning to phrases such as "A_n approaches A."

But make no mistake about it. Archimedes stands supreme among Greek mathematicians. Although much of his work has immediate practical applications, "he was wholly and passionately committed to mathematics at its purest."[1]

The Romans captured Syracuse in 212 BCE when Archimedes was seventy-five years of age. Following the fall of the city, the old man stood lost in thought and staring at a diagram he had drawn in the sand. While he looked, a Roman soldier killed him with a sword.

In spite of the great accomplishments of Greek mathematics, it possesses a "static" quality. Princeton's Salomon Bochner says:

> Now, it is precisely such an incapacity to conceptualize the rate of change which separates Archimedes from Newton by a threshold which Archimedes never succeeded in crossing.[2]

In particular, Greek mathematics lacked the concepts of *variable* and *function* that allow one to appropriately describe motion. We will turn to these descriptions momentarily. But, first, we still face the problem of making sense of the notion of area bounded by an arbitrary curve as shown in the last sketch of figure 58.

Calculus, as it is taught in college, is mathematics of the real numbers. Normally, the associated mathematics of the complex numbers composes a course of a higher level. Real numbers, of course, do not move. The number 6, for example, never takes a stroll and wanders past the number 7 on its way to 8. Nevertheless, calculus contains many allusions to moving numbers. In fact, part of the subject's power arises from the correctness it assigns to intuitive results obtained from dynamic numerical thought processes.

Think of the symbol x as denoting an *arbitrary* real number. Thus, x represents any number, and you can think of it—when you want it to change values—as moving along the real line, taking different values as it occupies different positions. Now suppose that, associated with each value of x, there is another number y. If the value of y depends on the value of x according to some known rule so that as x changes, y changes in some prescribed fashion, then, as we have seen, we call x and y *variables* and the rule that determines y in terms of x is called a *function*. If, as before, we denote the function by the letter "f," then we describe the statement "y is a function of x" by the notion $y = f(x)$. One function that we have encountered before is the function given by the quadratic equation $y = x^2$. Here the "rule" defining the function is the squaring process: y is determined by squaring x. The graph of the equation $y = x^2$, which has already been discussed, appears again in figure 60.

Similarly, one can—at least in theory—construct the graph of an arbitrary function $y = f(x)$. An example of such a graph appears also in figure 60. You will notice that the region bounded by the graph of the function and the two vertical lines represent a figure similar to a piece of the region of figure 60 whose area has not yet been determined. In the last sketch in figure 60, this region has been recopied from figure 58 and then decomposed into several pieces. The region in the middle, denoted by R_1, is a rectangle and we know how to compute its area. The region denoted by R_2 is exactly the region determined by the graph of $y = f(x)$ and the vertical lines at a and b. The other regions, R_3, R_4, and R_5, are simply translations and rotations of regions of the type given by the graph of $y = g(x)$, and appropriate vertical lines, where g is some other function.

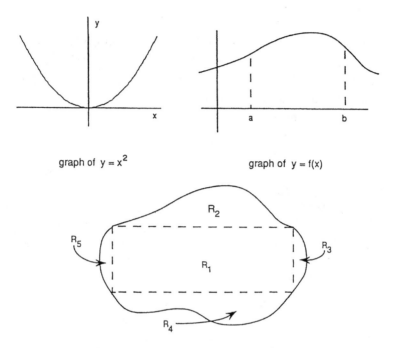

Figure 60: Decomposition of area

This results in the following mathematical fact that can be rigorously established: "most" regions bounded by curved lines, as shown in figure 60, can be decomposed into a finite number of regions of the type bounded by the graph of a function $y = f(x)$, the x axis, and appropriate vertical lines. Consequently, in order to determine the area enclosed by the more general region, it is sufficient to determine the area bounded by the graph of $y = f(x)$ and the two vertical lines.

Therefore, the problem of determining the area of a region bounded by curved lines reduces to the problem of determining the area formed by the graph of a function, the x axis, and two vertical lines. And we find this area as would Archimedes, had he available to him proper notation and the notion of a function. We use his method of exhaustion.

THE INTEGRAL

Figure 61 shows once more the graph of $y = f(x)$ and the two vertical lines drawn at the points $x = a$ and $x = b$. The symbol R denotes the region bounded by the graph, the two vertical lines, and the x axis. Let A denote the (as yet undefined) area of the region R. We want to simultaneously define and calculate the number A.

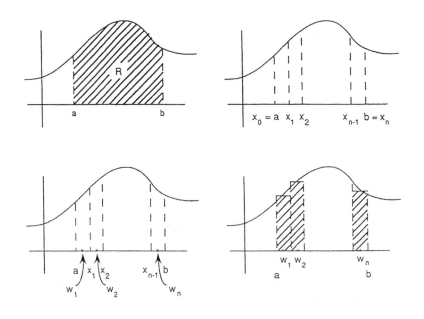

Figure 61: Area under a curve

In the second sketch of figure 61, the region R has been broken into "vertical strips" by the insertion of points on the x axis between a and b and the drawing of vertical lines through these points. Traditionally, these points are labeled $x_0, x_1, x_2, \ldots, x_n$, where $x_0 = a$ and $x_n = b$. These $n + 1$ points of division yield n strips into which R is divided. As figure 61 shows, the points of division are equally spaced so that each of the vertical strips has the same width.

Now, let us select a point within the base of each strip. That is, we select a point between x_0 and x_1, another between x_1 and x_2, and so on until we finally select a point between x_{n-1} and x_n. We denote these new points by w_1, $w_2, \ldots, w_n$. In the third sketch of figure 61 these new points are shown where, for convenience, each as been chosen in the middle of the base of its vertical strip.

In the fourth sketch of figure 61, shaded rectangles have been drawn by "squaring off" the tops of the vertical strips. You should observe that the height of each rectangle is the distance from the appropriate $w_1, w_2, \ldots, w_n$ to the graph of the function f. Therefore, these heights are exactly the values $f(w_1), f(w_2), \ldots, f(w_n)$. Thus $f(w_1)(x_1 - x_0)$, being the product of the height and the base, represents the area of the first x_n shaded rectangle. Similarly, $f(w_2)(x_2 - x_1), f(w_3)(x_3 - x_2), \ldots, f(w_n)(x_n - x_{n-1})$ are the respective values of the areas of the second shaded rectangle, and the third shaded rectangle, $\ldots$, and the nth shaded rectangle.

Let A_n denote the sum of the areas of the shaded rectangles. So,

(D) $$A_n = f(w_1)(x_1 - x_0) + f(w_2)(x_2 - x_1) + \ldots + f(w_n)(x_n - x_{n-1}).$$

Now, look carefully at figure 61 and notice that *the shaded rectangles closely approximate the* region R. Moreover, the approximation will become even better if we draw a new sketch and take more (and thinner) vertical strips.

This observation is analogous to the one made by Archimedes when he noticed he could approximate even more closely the region enclosed by a circle by filling it with polygons of an increasing number of sides. Here, in order to better the approximation, we must increase the number of vertical strips. And we do this by increasing the number of points of division, $x_1, x_2, \ldots, x_n$. Clearly, this is accomplished by letting n increase without bound. As the mathematicians say: "we let n tend to infinity." As this happens, A_n approaches more and more closely the unknown number A, which represents the area of the region R. In modern notation,

(E) $$A = \lim_{n \to \infty} A_n.$$

and we say: "the limit as n tends to infinity of A_n is A." (Notice that "infinity" has a different meaning here than in chapter 2, where we discussed infinite sets. This new notion is a dynamic concept while the earlier was more or less static in nature. Intuitively, to say that "n tends to infinity" means simply that n increases without bound, that is, that it ultimately exceeds any fixed number given in advance.)

Equation (E) then both *defines* and *calculates* for us the number A, at least in the cases for which the limit of A_n actually exists. (An example where the limit fails to exist is $A_n = n^2$. Here A_n becomes arbitrarily large, as n increases without bound. Thus A_n does not approach any fixed real number.) The early practitioners of calculus—people like Newton and Leibniz— proceeded, as did Archimedes, intuitively and their intuition was sufficiently keen that they obtained correct results. Only later were mathematicians such as Cauchy and Weierstrass able to replace the intuition with rigor and assign to these notions and procedures the label of "mathematical truth." And part of what students study in calculus today—or should study—consists of the careful examination of exactly what is intuition and what is truth.

For the moment, we will defer a detailed discussion of the notion of limit expressed in equation (E) and proceed intuitively. If, for example, we calculate A_n for a particular function f and it turns out to be

$$A_n = 6 - 2/n$$

then clearly $A_n = 6$. (As n becomes arbitrarily large, $\dfrac{2}{n}$ approaches 0.) On the other hand, if

$$A_n = n^2 + 1$$

then $A = \lim_{n \to \infty} A_n$ fails to exist because A_n increases without bound as n increases. But in the case of

$$A_n = 1 + \frac{1}{4} + \frac{1}{9} + \frac{1}{16} + \ldots + \frac{1}{n^2}$$

our intuition fails and whether A_n has a limit or not is uncertain. (It is known in this case that $\lim_{n \to \infty} A_n = \dfrac{\pi^2}{6}$. But this is by no means obvious.)

Much of the notation used in modern calculus owes its conception to the work of Leibniz. And one of the most useful symbols is that which represents the notion of the *integral*. The integral is written as

$$\int_a^b f(x)\,dx$$

and represents, for the function f, the procedure summarized in equations (D) and (E). Thus, the above symbol—which is read "the integral from a to b of *f* of *x*, dee *x*"—stands for the number *A*, which denotes the area of region *R*. The function value, $f(x)$, is called the *integrand*. Consequently,

(*F*)
$$\lim_{n \to \infty} A_n = \int_a^b f(x)\,dx$$

whenever the limit on the left-hand side of (*F*) exists.

Thus the integral represents a complicated procedure and, within it, are several ideas: a function, an interval, a subdivision, the selection of intermediate points, the formation of a certain sum, and the taking of a limit. The usefulness of calculus—and indeed, of all mathematics—results partially from the fact that, by developing rules for manipulating *symbols*, you develop procedures for manipulating complex *ideas*. Often, one can uncover deep truths through purely formal, manipulative means.

You should notice the suggestiveness of the integral notation. If you think of the "integral sign" as being "a long *S*" with *S* standing for "sum," then the notation becomes "the sum, from *a* to *b*, of *f(x)* times *dx*." Now think of $f(x)$ as the height of one of the thin vertical regions in the division of *R*. And think of *dx* as being the width of that thin rectangle. The notation then— with the limit understood—becomes a mnemonic for the entire area-defining process.

The early practitioners of calculus thought of *dx* as representing a width that, while being arbitrarily small, still did not equal zero. Such a quantity they called an *infinitesimal*.

Using modern methods one can *prove* that the number defined by (*E*), with A_n given by (*D*), exists whenever the function *f* is continuous, that is, whenever the graph of $y = f(x)$ has no breaks or jumps. Moreover, when *f* is continuous, the number *A* exists no matter how the points of division or the intermediate points are selected so long as the base of each rectangle shrinks to zero as the number of points of division tend to infinity. Such functions are said to be *integrable*.

And, fortunately, the number *A* given by (*E*) agrees with what we already know whenever the region *R* turns out to be a triangle or a rectangle or a circle. For example, consider the function given by $f(x) = \sqrt{r^2 - x^2}$. In this case the graph of $y = f(x)$ turns out to be a semicircle of radius *r* and center at the origin of the *x-y* plane. If we now apply the above procedure to this function, choosing $a = -r$, and $b = r$, we will obtain $A = \frac{1}{2}\pi r^2$, which agrees with the value Archimedes obtained all those centuries ago with his conceptually similar, but technically different, method.

I say we obtain this value and, in fact, we will obtain it provided the calculations can be done correctly in equation (D) and the limit can be evaluated in equation (E). Unfortunately, this cannot be easily accomplished. In fact, even for very simple functions, equations (D) and especially (E) lead to technical difficulties. Let's look at an example. Consider the now familiar function $f(x) = x^2$. Let's try to apply the procedure using $a = 0$ and $b = 1$. Thus, we want to calculate the area of the region R bounded by the graph of the equation $y = x^2$, the x axis, and the vertical lines at $x = 0$ and $x = 1$. This region is shown in figure 62. You will notice that the "vertical line" at $x = 0$ has degenerated, for this particular example, into a single point. In the integral notation, we want to compute

$$A = \int_a^b f(x)\,dx.$$

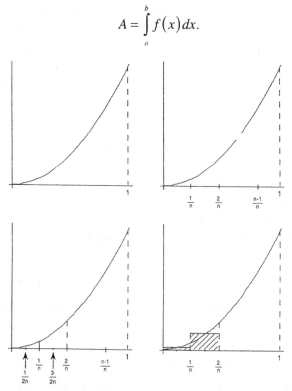

Figure 62: Approximation of area

In this case, the $n + 1$ points of subdivision become the points 0, $1/n$, $2/n$, ..., 1 and the points $w_1, w_2, \ldots, w_n$, become $1/2n$, $3/2n$, ..., $\dfrac{2n-1}{2n}$. (These values, being the midpoints of each of the intervals of subdivision, are obtained by adding the values of the endpoints and dividing by 2.) These points are shown in the second and third sketches of figure 62. When we substitute these values into equation (D) we obtain:

$$A_n = \left(\frac{1}{2n}\right)^2\left(\frac{1}{n}\right) + \left(\frac{3}{2n}\right)^2\left(\frac{1}{n}\right) + \ldots + \left(\frac{2n-1}{2n}\right)^2\left(\frac{1}{n}\right).$$

Although this expression can be simplified somewhat, it is not at all clear what value A_n approaches as n tends to infinity. Indeed, it is not clear from equation (F) that the limit of A_n, as n tends to infinity, even exists.

Fortunately, calculus allows us to compute the number A given by equation (E)—for this example and for many others—in a different manner. That is, there exists a way to compute

$$A = \int_a^b f(x)\,dx$$

other than by appeal to the definition that requires the evaluation of an often complicated limit. And it is exactly the existence of this "different manner" that gives calculus its power. Indeed, the existence and the simplicity of this alternative method defines the essence of the subject called *calculus*. The method itself is described in a theorem called the *fundamental theorem of calculus*. The theorem was proved by Newton and, independently, by Leibniz.

As we shall see, the fundamental theorem connects, in an astonishing manner, two apparently unrelated geometric notions: the area under a curve and the tangent line to the curve at an arbitrary point. But, for the moment, we will turn away from the "static" mathematics associated with area and tangent lines and look at things that move. We need to discuss the notion of the derivative of a function. And, following Isaac Newton, we will come to the derivative by examining the motion of a particle.

THE DERIVATIVE

Consider a particle moving along a line. Let us suppose that, at any time t, the particle has traveled a distance s from some given starting point. Thus, we are assuming that, for the particle, the distance s can be expressed as some known function of time t. Suppose we are given

$$s = f(t).$$

Figure 63 shows the particle as a point on its line of motion and indicates the distance s at some particular time. The second sketch in figure 63 represents the graph of the equation $s = f(t)$ in the t-s coordinate plane. The graph of $s = f(t)$ will ordinarily be some curve as figure 63 indicates. However, you must understand that *the particle does not move on this curve but rather on the straight line of the first sketch*. The graph of $s = f(t)$ describes the relationship between time and distance. When, for example, the graph is rising, then the distance *increases* with time. A falling graph indicates the particle has turned around and the distance from the starting point decreases. Notice that no graph exists for negative values of t since this would have no physical meaning.

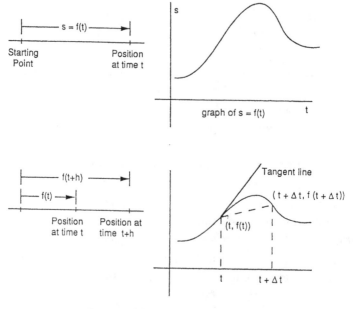

Figure 63: Motion as a graph

Newton's famous laws are laws of *motion*, and, in order to express them, Newton needed mathematics sufficient to describe *velocity* and the rate of change of velocity, which is called *acceleration*. But what he needed first was to assign some precise meaning to the concept of *instantaneous velocity*. The particle of figure 63 moves according to the function $s = f(t)$ and, presumably, speeds up and slows down in an erratic manner. What do we mean when we say that it has a certain velocity at a particular instant?

Let's suppose we want to know the particle's velocity at time t. Think of a slightly advanced instant, $t + h$. At time t, the particle has gone the distance $f(t)$. At time $t + h$, the particle has traveled the distance $f(t + h)$. (See sketch 3 of figure 63.) Consequently, in the interval of time between t and $t + h$, the particle has traveled a distance of $f(t + h) - f(t)$. If we divide this distance by the elapsed time, h, we will obtain the "average velocity" of the particle in the time period t to $t + h$.

Thus

$$\text{average velocity} = \frac{f(t+h) - f(t)}{h}.$$

A more suggestive notion arises when we denote the numerator of the above fraction by the symbol Δs and we replace the letter h by the symbol Δt. We think of "Δ" as denoting "difference" or "change" so that

$$\Delta s = f(t + \Delta t) - f(t)$$

represents the change in distance traveled by the particle as t changes from t to $t + \Delta t$. (The Δs notation represents another exception to our usual convention that juxtaposition of symbols indicates multiplication.) With this notation the expression for average velocity of the particle in the time interval t to $t + \Delta t$ becomes

$$\frac{\Delta s}{\Delta t} = \frac{f(t + \Delta t) - f(t)}{\Delta t}.$$

Newton's idea was to consider the behavior of the average velocity as the time interval became smaller and smaller. The limiting value of the average velocity as the time interval t to $t + \Delta t$ shrinks to zero is what Newton *defined* as the instantaneous velocity at time t. Let's denote this velocity by $v(t)$. Thus,

(G)
$$v(t) = \lim_{\Delta t \to 0} \frac{f(t + \Delta t) - f(t)}{\Delta t}$$

provided, of course, that this limit exists.

In the fourth sketch of figure 63, points are drawn on the graph of $s = f(t)$ corresponding to time t and $t + \Delta t$. Also shown is the so-called *secant line* joining the points $(t, f(t))$ and $(t + \Delta t, f(t + \Delta t))$. Notice that, as Δt approaches 0, this secant line seems to approach the *tangent line* to the graph of $s = f(t)$ at the point $(t, f(t))$.

Formally, the expression for $v(t)$ given by (G) "looks like" the meaningless expression 0/0 since both the numerator and the denominator on the right-hand side tend to zero with Δt. But looks can be deceiving and in practice things work out differently. Let's consider a simple example.

Suppose the particle moves according to the rule $s = t^2$. Then,

$$\frac{\Delta s}{\Delta t} = \frac{(t + \Delta t)^2 - t^2}{\Delta t}.$$

Simplification yields

$$\frac{\Delta s}{\Delta t} = \frac{t^2 + 2t\Delta t + (\Delta t)^2 - t^2}{\Delta t}$$

$$= \frac{2t\Delta t + (\Delta t)^2}{\Delta t}.$$

For all values of Δt *different from zero* we may *divide* the numerator and the denominator of the above fraction by t. This gives

$$\frac{\Delta s}{\Delta t} = 2t + \Delta t.$$

Now, we take the limit as Δt approaches zero and we obtain

$$v(t) = \lim_{\Delta t \to 0} \frac{\Delta s}{\Delta t}$$

$$= \lim_{\Delta t \to 0} (2t + \Delta t)$$

$$= 2t.$$

The appropriate cancellation of the troublesome Δt from the denominator removed the difficulties associated with the meaningless symbol "0/0."

Of course, our calculation was purely formal and we have not attempted to give a precise meaning to the "limit process." Neither, in fact, did Newton or Leibniz. (The modern definition of limit did not, in fact, appear until Weierstrass formulated it around 1880.)

In the general case given by equation (*G*), the limiting value of the right-hand side, when it exists, is called the *derivative* of the function *f* with respect to *t*. There are two common symbols for the derivative: f' and df/dt $\left(\text{or } \dfrac{ds}{dt}\right)$.

Therefore, by definition, the derivative is

$$\frac{ds}{dt} = f'(t) = \lim_{\Delta t \to 0} \frac{\Delta s}{\Delta t} = \lim_{\Delta t \to 0} \frac{f(t + \Delta t) - f(t)}{\Delta t}.$$

The velocity, *v(t)*, is then the derivative of the position function *s(t)*. And, in our example, we have shown that the derivative of $s(t) = t^2$ is $s'(t) = 2t$. Or, in the "*d* notation,"

$$\frac{d(t^2)}{dt} = 2t.$$

Newton's second law of motion involved, as he expressed it, the "rate of change of momentum." Since momentum equals mass times velocity and, to Newton, moving bodies had constant mass, "rate of change of momentum" involved rate of change of velocity, that is, *acceleration*.

You deal with acceleration exactly as you do velocity and, consequently, acceleration becomes the *derivative of the velocity function*. So, if we denote the acceleration at time *t* by *a(t)*, we have

$$a(t) = v'(t).$$

Since *v(t)* is itself the derivative of *s(t)*, *a(t)* is the *second derivative of s(t)*. This is written:

$$a(t) = s''(t)$$

or

$$a(t) = \frac{d^2 s}{dt^2}.$$

In the case of our example $s(t) = t^2$, it can be shown that

$$a(t) = s''(t) = 2.$$

Thus a particle moving according to the equation $s = t^2$ has *constant acceleration*, like a falling stone.

Newton had motion on his mind so he considered functions that expressed position as a function of time. But the concepts of limit and derivative carry over identically to more general functions expressed in the ordinary $y = f(x)$ notation. You compute the limit of the "difference quotient" exactly as in the case of equation (G) and you are led to the "derivative of y with respect to x." Thus

(H)
$$\frac{\Delta y}{\Delta x} = \frac{f(x + \Delta x) - f(x)}{\Delta x}$$

and

(I)
$$\frac{dy}{dx} = f'(x) = \lim_{\Delta x \to 0} \frac{f(x + \Delta x) - f(x)}{\Delta x}.$$

And, by repeating the process, we can define successive derivatives $f''(x), f'''(x)$, and so on. Our example, when t and s are replaced by x and y shows:

$$f(x) = x^2$$
$$f'(x) = 2x$$

and

$$f''(x) = 2.$$

One more differentiation of this function gives

$$f'''(x) = 0.$$

The proofs of these last two statements are easy. All we need do is write definition (I) in the form

$$g'(x) = \lim_{\Delta x \to 0} \frac{g(x + \Delta x) - g(x)}{\Delta x}$$

and apply it successively to the cases $g(x) = f'(x)$ and $g(x) = f''(x)$. In the first case we have

$$g'(x) = \lim_{\Delta x \to 0} \frac{f'(x + \Delta x) - f'(x)}{\Delta x}.$$

But, if $g(x) = f'(x)$ then $g'(x) = f''(x)$. So we have

$$f''(x) = \lim_{\Delta x \to 0} \frac{f'(x + \Delta x) - f'(x)}{\Delta x}.$$

In this example, $f'(x) = 2x$. So, the above equation becomes

$$f''(x) = \lim_{\Delta x \to 0} \frac{2(x + \Delta x) - 2x}{\Delta x}$$

$$= \lim_{\Delta x \to 0} \frac{2x + 2\Delta x - 2x}{\Delta x}$$

$$= \lim_{\Delta x \to 0} \frac{2\Delta x}{\Delta x}$$

$$= \lim_{\Delta x \to 0} 2$$

$$= 2.$$

In the last section, we introduced the integral by looking at the geometric concept of area. Here, we choose to look at the derivative from the point of view of the motion of a particle because of its historical significance and because of its familiarity. But we must now turn back to geometry and point out the connection between the notion of derivative and the notion of tangent to a curve.

Figure 64 essentially reproduces the fourth sketch of figure 63 for the case of a general function $y = f(x)$. The figure shows points on the graph having coordinates $(x, f(x))$ and $(x + \Delta x, f(x + \Delta x))$ and the secant line joining these points. It also shows the tangent line to the graph at the point $(x, f(x))$. Notice that the tangent of the angle θ of figure 64 equals the difference quotient given by equation (H). That is,

$$\tan \theta = \frac{\Delta y}{\Delta x}.$$

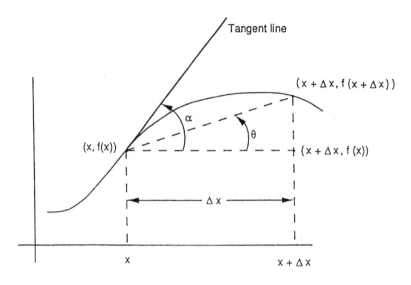

Figure 64: Tangent to a curve

The other angle indicated in the sketch, α, is the angle that the tangent line makes with the horizontal and is called "the angle of inclination of the tangent line." The tangent of this angle is called the *slope*.

Now look at figure 64 and imagine the quantity Δx as tending to zero. As Δx shrinks, the point $(x + \Delta x, f(x + \Delta x))$ slides down the graph of $y = f(x)$ heading toward the point $(x, f(x))$. As the point moves, so does the secant line. And, as it moves, it comes closer and closer to the tangent line at $(x, f(x))$ and θ simultaneously approaches the angle α. Thus, the limit of the difference quotient becomes the tangent of angle α. But this limit is the derivative as given by equation (*I*). Hence, *the derivative,* f '(x), *equals the slope of the tangent to the graph of* y = f(x) *at the point* (x, y).

What actually happens here is that you define the tangent line by this process so that curves possessing tangent lines are exactly the graphs of differentiable functions. Consequently, differentiability implies a kind of *smoothness* and you would expect a function *not* to be differentiable at points where their graphs had corners or sharp points. And, although this can be shown to be true, there are many subtleties associated with the set of points at which a function fails to be differentiable. The early mathematicians were vague and uncertain about these matters and, in some cases, plainly incorrect. We will return to this when we look more carefully at the notion of limit.

But now we need to examine the result that makes calculus work. Before Newton and Leibniz, "calculus" consisted of a jagged collection of ideas and techniques scattered over the mathematical analysis landscape like stones. Newton and Leibniz, independently of one another, discovered the keystone that would bind these ideas together into a great arch—a gateway to truths of mathematics and of nature. The keystone is called the *fundamental theorem of calculus*.

THE FUNDAMENTAL THEOREM

On the face of it, derivatives and integrals seem to be unrelated. The derivative arises through the taking of a certain limit of the difference quotient as equation (*I*) shows. The integral comes from a complicated process requiring the partitioning of an interval, the formation of a particular sum, and the taking of a limit as described in equations (*D*) and (*F*). Geometrically, the derivative, $f'(x)$, represents the slope of the tangent to the curve $y = f(x)$ while the integral

$$\int_a^b f(x)dx$$

stands for the area under the same curve between a and b. There is no a priori reason to believe the two concepts are related. But they are. And the nature of the relationship constitutes the most important and elegant theorem in calculus.

First, let's notice that the integral written above, being an area, represents some fixed number. That is, the *integral* of $f(x)$ from a to b is neither a variable nor a function but is a number. Earlier, we have denoted this number by the letter A.

Consequently, the integral does not depend on the "x" that appears inside. (Equation (*D*) shows that there is no x in the expression A_n whose limit provides the definition of $\int_a^b f(x)dx$ in equation (*F*).) Mathematicians, for this reason, call the x a "dummy variable." We could, in fact, replace the x with any other letter (except a, b, f, or d) and the integral would be unchanged. For example,

$$\int_a^b f(x)\,dx = \int_a^b f(s)\,ds = \int_a^b f(w)\,dw = \int_a^b f(t)\,dt.$$

However, we can convert the integral into a function of x by replacing the "b" in the upper limit by "x." That is, we think once more of the area under the graph of $y = f(x)$ as shown in figure 65. But now, we consider only that portion of the area bounded on the left by a vertical line at the point a and on the right by a vertical line at a *variable* point x. This new area (as shown in figure 65) certainly depends on x and, therefore, determines a new function $F(x)$. But we know an explicit "formula" for $F(x)$. Since it represents an area of the type that led us to the integral, $F(x)$ can also be written as an integral; namely, as the integral of $f(s)$ between a and x:

(J) $$F(x) = \int_a^x f(s)\,ds.$$

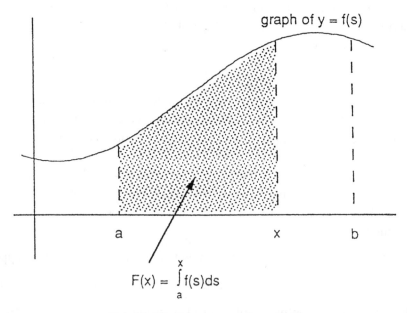

Figure 65: Area as an integral

(Here, we must not use x as the inside "dummy variable" in order to avoid confusion with the legitimate variable of the same name in the upper limit.)

Let's notice that $F(a) = 0$ since this represents the area under the graph between a and b. Also, $F(b) = A$, the original area described in equation (E).

Before we can state the fundamental theorem, we need the concept of *antiderivative*. In our earlier example (with different notation), we showed that if $f(x) = x^2$, then $f'(x) = 2x$. Thus, if we let $g(x) = 2x$, then the function g is *the derivative* of the function f. And we say that f is *an antiderivative* of the function g.

Notice that I have used the articles "the" and "an" in the previous sentence. We speak of "the" derivative of a given function f because f', if it exists at all, would be unique. When you take the limit of the right-hand side of equation (I) as x tends to zero, you get, at most, one answer. But a different situation prevails in the case of antiderivatives. Antiderivatives are *not* unique. If g is an antiderivative of f, then so is the function $g(x) + c$ where c is any constant. Hence, if a function has an antiderivative, it has infinitely many with any pair of antiderivatives differing by a constant. Let's examine this situation briefly.

The derivative of any constant function equals zero. This fact follows easily from the definition given in equation (I). Consider the function $f(x) = c$ where c is a constant. Substitute this function into the right-hand side of equation (I). The two terms in the numerator of the difference quotient are both equal to c because the function, being constant, always equals c. Consequently, the numerator equals 0 and the difference quotient becomes $0/\Delta x$, which itself equals 0 for all $\Delta x \neq 0$. Hence, the derivative, being the limit of something that identically equals zero, is itself zero.

More precisely, the argument is: let $f(x) = c$. Then

$$f'(x) = \lim_{\Delta x \to 0} \frac{f(x + \Delta x) - f(x)}{\Delta x}$$

$$= \lim_{\Delta x \to 0} \frac{c - c}{\Delta x}$$

$$= \lim_{\Delta x \to 0} \frac{0}{\Delta x}$$

$$= \lim_{\Delta x \to 0} 0$$

$$= 0.$$

The result also follows easily on intuitive grounds. If $f(x) = c$, then the graph of $y = f(x)$ becomes a horizontal straight line of height c above the *x* axis. Therefore, the tangent to the graph is everywhere this same straight line. Hence, the graph everywhere has zero slope. But this slope equals $f'(x)$. Thus $f'(x) = 0$ for all *x*.

Conversely, one can prove—not so easily—that if $f'(x) = 0$ for all values of *x* for which $f(x)$ is defined, then $f(x) = c$, for some constant *c*. That is, a function whose derivative identically equals zero reduces to a constant. This result also follows intuitively: if $f'(x) = 0$ for all *x*, then the tangent line to the graph of $y = f(x)$ always has zero slope. Consequently, the graph can neither rise nor fall. So, the graph of $y = f(x)$ is flat. Hence, $f(x) = c$ for some constant *c*. (A precise proof of this requires a result known as the mean value theorem.)

Let's return to the question of nonuniqueness of antiderivatives. Suppose, as before, that *g* is an antiderivative of *f*. Hence, $g'x) = f(x)$. Let *c* be any constant. Let $h(x) = g(x) + c$. Then,

$$h'(x) = \frac{d}{dx}\left(g(x) + c\right)$$
$$= \frac{d}{dx}g(x) + \frac{d(c)}{dx}.$$

(The second step follows because "the derivative of a sum equals the sum of the deriatives." This fact can be easily proved from the definition of derivative and is a partial example of a property mathematicians call "linearity.") But, we know the derivative of a constant equals zero. Thus

$$\frac{d(c)}{dx} = 0.$$

Hence,

$$h'(x) = g'(x) = f(x).$$

Therefore, *h* is also an antiderivative of *g*.

Now, let's suppose that *h* and *g* are any two antiderivatives of *f*. Thus, $h'(x) = g'(x) = f(x)$. Consider the function *F* defined by

$$F(x) = h(x) - g(x).$$

Then

$$F'(x) = h'(g) - g'(x)$$
$$= f(x) - f(x)$$
$$= 0.$$

Consequently, $F'(x)$ is identically zero. And we know that such a function reduces to a constant. Hence, there exists a c such that $F(x) = c$. Thus

$$h(x) - g(x) = c$$

and

$$h(x) = g(x) + c.$$

So, we now have settled the issue of the article "an" when we speak of antiderivatives. The facts are these: whenever a function has an antiderivative, it has infinitely many, but any two antiderivatives differ by a constant.

Let's return to the situation described by equation (J). Our assumptions are that we have a"nice" function defined at least for all values of x between a and b. Let $F(x)$ be defined by J, that is

$$F(x) = \int_a^x f(s)\,ds.$$

The *fundamental theorem of calculus* says two things:
(i) F is an antiderivative of f, that is, $F'(x) = f(x)$

and

(ii) if H is any antiderivative of f, then

$$\int_a^b f(s)\,ds = H(b) - H(a).$$

Notice that (i) can be written as

$$\frac{d}{dx}\left(\int_a^x f(s)\,ds\right) = f(x).$$

And, if we take F as one of the choices for H in (ii), we have

$$\int_a^b F'(s)\,ds = F(b) - F(a).$$

Consequently, the fundamental theorem tells us that differentiation and integration are *reciprocal* operations of one another in the sense that, if you differentiate an integral with respect to its upper limit, you get the integrand evaluated at the upper limit and if you integrate the derivative of a function you recover the function, evaluated between the limits of integration.

These facts, which are just restatements of the previous two equations, brought differentiation and integration together and cemented the grab-bag collection of earlier results into the mathematical arch that is now called calculus. Conclusion (ii) of the fundamental theorem provides the "different manner" of evaluating integrals I mentioned earlier.

Recall our difficulty in trying to evaluate the seemingly simple integral

$$\int_0^1 x^2 dx.$$

Part (ii) of the fundamental theorem gives the value immediately provided you know any antiderivative of the function $f(x) = x^2$. For if H is any such antiderivative then (ii) says

$$\int_0^1 x^2 dx = H(1) - H(0).$$

Calculus students spend considerable time learning techniques for computing derivatives and antiderivatives of many basic functions. And the simplest functions with which they deal are single-term polynomials.

You will recall that we showed

$$\frac{d(x^2)}{dx} = 2x.$$

Similarly, one can show that

$$\frac{d(x^3)}{dx} = 3x^2$$

and

$$\frac{d(x^4)}{dx} = 4x^3.$$

This pattern leads you to *conjecture* that, for any positive integer n,

$$\frac{d(x^n)}{dx} = nx^{n-1}.$$

The result is, in fact, true and the proof depends only on two results described earlier: the binomial theorem and mathematical induction.

Whenever you have a formula expressing a derivative, you have—if you read it in reverse order—a formula for an antiderivative. For example, the last formula above says that the derivative of x^n is nx^{n-1}. It also says that the antiderivative of nx^{n-1} is x^n. If we rearrange the constant n appropriately in this last statement and then replace n by $n + 1$, we obtain

$$\frac{x^{n+1}}{n+1} \text{ is an antiderivative of } x^n.$$

We should note, in passing, that this last result requires that $n \neq -1$ since, otherwise, we would be dividing by zero. In the case of $n = 2$, the statement tells us

$$\frac{x^3}{3} \text{ is an antiderivative of } x^2.$$

Now our integral falls immediately to the fundamental theorem:

$$\int_0^1 x^2 dx = H(1) - H(0).$$

Take $H(x) = x^3/3$. Then

$$\int_0^1 x^2 dx = \frac{1}{3} - \frac{0}{3} = \frac{1}{3}.$$

I do not intend to say anything about the proof of the fundamental theorem of calculus other than that a quite plausible argument can be given even for beginning students. Of course, the proof—and even a precise statement of the theorem—depends on clarifying the assumption I have hidden under the phrase "nice function." Normally, one assumes the function in the fundamental theory to be "continuous," which means intuitively that its graph has no jumps or breaks. The theorem actually holds under weaker conditions. But continuity is nice and will suffice.

Before we can say anything more certain about continuity we need to examine closely the central notion on which it is based. This notion also forms the key idea in the definition of both the integral and the derivative. It, in fact, is the central notion in all of calculus although its precise statement eluded mathematicians as great as Newton and Leibniz all their lives.

From the beginning, calculus was applied to "the architecture of the universe,"[3] but the failure of early mathematics to come to grips with the subject's central notion made its acceptance more a matter of faith than of mathematical logic. Voltaire described calculus as

> the art of numbering and measuring a Thing whose existence cannot be conceived.[4]

And Bishop George Berkeley, raging against the attempts of Newton and Leibniz to deal intuitively with the central notion through "fluxions" and "infinitesimals," said bitterly:

> They are neither finite quantities, nor quantities infinitely small, nor yet nothing. May we not call them the ghosts of departed quantities.[5]

The central notion that eluded them all was the notion of *limit*.

THE LIMIT

Calculus lives in the mathematical world and consequently is composed of ideas. But the notions have structure and together they give the subject form and shape. Think of the ideas as stones that when fitted together form a great arch (see figure 66). The two prongs of the arch have names: *integral calculus* on the left and *differential calculus* on the right. Integral calculus contains the mathematics that grew from attempts to understand and to compute the area enclosed by certain classes of curved lines. As we have seen, early contributions to this branch of calculus were made by Archimedes of Syracuse around 200 BCE. The ideas of differential calculus came essentially from the efforts of mathematicians to understand the concept of instantaneous velocity and the associated idea of the notion of tangent line to a curve. We know that Pierre Fermat contributed to the development of this part of the subject. But so did other seventeenth-century mathematicians, like John Wallis and Blaise Pascal.

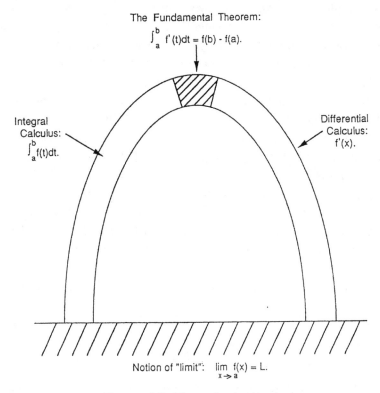

Figure 66: The calculus arch

But the two mathematicians who justly receive the most credit for the creation of calculus are Isaac Newton (1642–1727) in England and Gottfried Leibniz (1646–1716) in Germany. These men independently—and almost simultaneously—showed that the apparently unrelated ideas of area on the one side and of tangent line on the other are, in fact, closely linked. They established the connection between the two by formulating what now is called the *fundamental theorem of calculus.* As figure 66 shows, the fundamental theorem forms the crownstone of the calculus arch. It is the existence of this theorem that allows the joining of the two sets of ideas and establishes the subject called calculus.

The value of the Newton-Leibniz accomplishment cannot be overestimated. All of the ideas of the continent of the mathematical world called *analysis* spring from calculus and, thus, from the work of these two men. Included are areas—both classical and modern—such as differential equations, complex variables, and approximation theory. To be sure, these subjects and the others that belong under the analysis rubric have now grown far beyond any concepts set down by Newton and Leibniz, but the original flavor remains. The sap that flows this day through all of analysis is the juice of calculus.

Moreover, the kind of mathematics, mainly used in the construction of mathematical models, is the mathematics of analysis and, hence, that of calculus. Thus, the calculus arch not only supports all of analysis but also most of the mathematics that is *applicable* to the real world. Much of the magic of mathematics is calculus magic.

But, while mathematical analysis lives above the calculus arch, there is something further down. The two prongs themselves stand on common ground. The integral and the derivative are each defined in terms of a certain type of limiting process. Recall

$$\int_a^b f(x)\,dx = \lim_{n \to \infty} A_n$$

and

$$f'(x) = \lim_{\Delta x \to 0} D(\Delta x)$$

where A_n is defined by (D) and $D(\Delta x)$ represents the difference quotient on the right side of equation (1).

Both the integral and the derivative are, therefore, defined in terms of the concept of limit. Consequently, the notion of *limit* is more fundamental than either the concept of *integral* or of *derivative*. The limit notion, then, joins the prongs of the arch at the bottom just as the fundamental theorem of calculus joins them at the top. The idea of *limit* is the bedrock on which calculus itself stands.

Ernest Hemingway said that if you want to understand bullfighting you must sooner or later come to Madrid. Similarly, if you want to understand calculus, you must sooner or later come to the notion of limit. Let's come to it now.

We will first consider limits of the form

(*K*)
$$\lim_{x \to a} g(x) = L.$$

A change of notation will reduce the limit on the right-hand side of equation (*I*) to a limit of the form given by (*K*) so that, by making (*K*) precise, we will simultaneously give precise meaning to the notion of the derivative of a function.

Here's one way to handle the "change of notation." In equation (*I*), *x* is fixed and Δx is the variable that is tending to zero. First, replace *x* by some other symbol, say the letter *c*. Then (*I*) becomes

$$f'(c) = \lim_{\Delta x \to 0} \frac{f(c + \Delta x) - f(c)}{\Delta x}.$$

Now we can replace the variable Δx by simply *x* and obtain

$$f'(c) = \lim_{x \to 0} \frac{f(c + x) - f(c)}{x}.$$

In the right-hand side of this expression, the variable now is *x*. If we write

$$g(x) = \frac{f(c + x) - f(c)}{x}$$

and

$$f'(c) = L$$

then the expression defining $f'(c)$ becomes simply equation (*K*), where *a* = 0. The notation in (*K*) is read:

"The limit as *x* approaches *a* of *g*(*x*) equals *L*."

Our task is to give a precise and rigorous meaning to (*K*). And, since this accomplishment elude dmathematicians of the genius of Newton and Leibniz, we do not expect it to be easy. (It is not easy and the remainder of this section constitutes the most subtle and sophisticated material in this book. It will require reflection and, probably, several readings. But stay with it and one day you will see it in a flash of light. And from that moment on, it will be forever yours.)

The notion of *limit*, expressed in (*K*), forms the base of all of calculus and constitutes the subject's most difficult idea. While you can become *literate* in calculus without understanding (*K*), you cannot become *knowledgeable*. To *know* calculus requires that you *know* the notion of limit. To "learn" calculus without learning this notion is like playing "tennis" without a net: easier, but not satisfying. And not tennis.

There are various ways to motivate the modern definition of (*K*). But all are defective in one way or another and to use them is merely to delay the intellectual pain associated with the definition. For grown-up people, it is best just to go at it directly—to take the sword and go in over the horns. Here's the definition. (The Greek letters epsilon (ε) and delta (δ) used in the definition may seem strange. But they are traditional and should not be avoided.)

(*L*) DEFINITION. We say that the "limit as *x* approaches *a* of *g*(*x*) is *L*" and we write

$$\lim_{x \to a} g(x) = L$$

if and only if for each $\varepsilon > 0$ there exists a number $\delta > 0$ such that

$$0 < |x - a| < \delta \text{ implies } |g(x) - L| < \varepsilon.$$

Notice, at the outset, that the definition of limit *says nothing whatsoever about the behavior of g*(*x*) *at x = a*. In fact, the inequality on the left-hand side of

$$0 < |x - a| < \delta$$

specifically prevents *x* from taking on the value *a*. Consequently, when we think of "*x* approaching *a*," we should think of "*x* approaching *a* but never being equal to *a*."

On the other hand, the definition allows the possibility that *g*(*x*) = *L* since there is no "greater than zero" symbol on the left of

$$|g(x) - L| < \varepsilon.$$

Notice also that if *g*(*x*) = *L* for all *x*, that is, *g* is a constant function, then

$$|g(x) - L| = |L - L| = 0 < \varepsilon.$$

So, for the constant function *g*(*x*) = *L*, the inequality

$$|g(x) - L| < \varepsilon$$

always holds.

In order to demonstrate (for an arbitrary function g) that

$$\lim_{x \to a} g(x) = L$$

the definition requires that for a given $\varepsilon > 0$, we *produce* $\delta > 0$ such that

$$0 < |x - a| < \delta \text{ implies } |g(x) - L| < \varepsilon.$$

Thus ε *is given in advance and* δ *must then be produced.*

The ε is a measure of *precision*. It is handed to us in advance. We must then produce another measure of precision, δ, such that $g(x)$ will be closer to L than ε whenever x is closer to a than δ but $x \neq a$. The gist is that, in order to show

$$\lim_{x \to a} g(x) = L$$

we must show that $g(x)$ can be made *arbitrarily* close to L by making x *sufficiently* close to a but not equal to a. The "arbitrary closeness" is measured by the $\varepsilon > 0$, which someone gives us. The "sufficient closeness" is measured by the $\delta > 0$, which we produce. Naturally, we expect the δ (which we must find) to depend on the ε (which is given to us). We expect that a "very small ε" will normally require a "very small δ."

Let's make one more observation about definition (L) before we come to an example:

> If there exists a single $\delta > 0$ for which definition (L) holds, then there exist infinitely many δ's that also satisfy the definition.

To see this, suppose we have found $\delta > 0$ that satisfies definition (L). Thus,

$$\text{for } \varepsilon > 0, \ 0 < |x - a| < \delta \text{ implies } |g(x) - L| < \varepsilon.$$

Now let δ_1 be any positive number less than δ, that is,

$$0 < \delta_1 < \delta.$$

Then

$$0 < |x - a| < \delta_1, \text{ implies } 0 < |x - a| < \delta$$

and, therefore

$$\left|g(x) - L\right| < \varepsilon.$$

So, δ_1 also satisfies definition (L). Clearly there are infinitely many choices for δ_1. (For example, $\delta_1 = \delta/2$ will do. Or $\delta_1 = \delta/3$.)

Here's an example that is intuitively clear but is as technically and conceptually difficult as anything encountered in freshman calculus.

EXAMPLE. Prove that $\lim\limits_{x \to 3} x^2 = 9$.

Intuitively, the result is clear. Obviously "x^2 is arbitrarily close to 9 provided x is sufficiently close to 3, but $x \neq 3$." We must do more to produce a *proof.*

Here's what must be done according to definition (L): we must prove that, for $\varepsilon > 0$, there exists a $\delta > 0$ such that

$$0 < \left|x - 3\right| < \delta \text{ implies } \left|x^2 - 9\right| < \varepsilon.$$

We begin the proof this way:

Let $\varepsilon > 0$. Choose $\delta = \_\_$. Then

$$0 < \left|x - 3\right| < \text{ implies } \left|x^2 - 9\right| \ldots < \varepsilon.$$

We must sooner or later make a specific choice for δ such that, whenever $0 < |x - 3| < \delta$, it follows that $|x^2 - 9| < \varepsilon$. Therefore, in the above scheme we must fill in the blank following the "$\delta =$" in such a way that our choice then allows us to manipulate the term $|x^2 - 9|$ in the space shown in the three dots to give a string of equalities and inequalities that will result finally in $|x^2 - 9| < \varepsilon$. Of course, we do not know in advance how to choose δ. This will come later—if we are clever or lucky or both—after we have done some manipulation with the term $|x^2 - 9|$. But it is better that we proceed, as the above scheme shows, with the unknown choice of δ clearly in front of us. For this is what proofs involving limits are all about: for a *given* $\varepsilon > 0$, you must choose an appropriate positive δ. Limit proofs should always begin with the statements:

Let $\varepsilon > 0$. Choose $\delta = \_\_$.

Somewhere, in the course of the proof, you must fill in the blank.

Let's begin anew by stating our task as a proper mathematical theorem and producing a proof. (At one stage we will need to use a result known as the triangle inequality. This says that for any real numbers a and b, $|a + b| \le |a| + |b|$.)

THEOREM.

$$\lim_{x \to 3} x^2 = 9.$$

PROOF. Let $\varepsilon > 0$. Choose $\delta =$_. Then $0 < |x - 3| < \delta$ implies

$$\begin{aligned}
\left|x^2 = 9\right| &= \left|(x - 3)(x + 3)\right| = \left|x - 3\right|\left|x + 3\right| \\
&= \left|x - 3\right|\left|x - 3 + 6\right| \\
&\le \left|x - 3\right|\left(\left|x - 3\right| + \left|6\right|\right) \\
&= \left|x - 3\right|\left(\left|x - 3\right| + 6\right) < \delta\left(\left|x - 3\right| + 6\right).
\end{aligned}$$

At this stage, let's *suppose* that $\delta \le 10$. Then it follows that

$$0 < |x - 3| < \delta \text{ implies } |x - 3| < 10.$$

Therefore

$$0 < |x - 3| < \delta \text{ implies } |x^2 - 9| < \delta(10 + 6)$$

or

$$|x^2 - 9| < 16 \cdot \delta.$$

Now, for the given $\varepsilon > 0$, let's choose $\delta \le \dfrac{\varepsilon}{16}$. Then we have

$$0 < |x - 3| < \delta \text{ implies}$$

$$|x^2 - 9| < 16 \cdot \delta \le 16(\varepsilon / 16) = \varepsilon.$$

Then

$$0 < |x - 3| < \delta \text{ implies } |x^2 - 9| < \varepsilon.$$

This implication is what we set out to prove; but we have essentially made two choices for δ:

$$\delta \le 10 \text{ and } \delta \le \varepsilon / 16.$$

But we can meet these conditions simultaneously by choosing δ to be the smaller of the two numbers 10 and $\varepsilon/16$. That is, we choose

$$\delta = \text{minimum } (10, \varepsilon / 16).$$

When you write this choice of δ in the blank space, the proof unfolds from left to right as easily as the Mikado's silk fan. Here's a recapitulation.

Let $\varepsilon > 0$. Choose $\delta = $ minimum $(10, \varepsilon/16)$. (Then $\delta \leq 10$ and $\delta \leq \varepsilon/16$.) Thus

$$0 < |x - 3| < \delta \text{ implies}$$

$$|(x^2 - 9) = |(x - 3)(x + 3)||$$

$$= |x - 3||x + 3|$$

$$< \delta|x + 3|$$

$$= \delta|x - 3 + 6|$$

$$\leq \delta(|x - 3| + |6|)$$

$$= \delta(|x - 3| + 6)$$

$$\leq \delta(10 + 6)$$

$$= 16\delta$$

$$\leq 16 \cdot \varepsilon/16$$

$$= \varepsilon.$$

Then,

$$0 < |x - 3| < \delta \text{ implies } |x^2 - 9| < \varepsilon$$

and the proof is complete.

You should notice that there is no particular reason for the preliminary choice $\delta \leq 10$ (except that it works). We might have, for example, chosen $\delta \leq 2$, at the preliminary stage. This would replace the earlier inequalities with:

$$|x^2 - 9| < \delta(|x - 3| + 6)$$

$$\leq \delta(2 + 6)$$

$$= 8 \cdot \delta$$

and then we would have had to choose $\delta \leq \varepsilon/8$ to force the final inequality.

$$|x^2 - 9| < \varepsilon.$$

In this case, the final choice of δ would have been

$$\delta = \text{minimum } (2, \varepsilon/8).$$

Infinitely many other choices for δ are also possible. But we only need one. Our choice of

$$\delta = \text{minimum } (10, \varepsilon/16)$$

works just fine.

Notice that the value of the limit itself, the number L, appears in definition (L). Consequently, we must "know" the limit in advance in order that we can use definition (L) to justify it. In our example, it was obvious that $L = 9$. In fact, if we write $h(x) = x^2$, then we see $L = 9 = h(3)$. That is—in this example—"the limit of the function $h(x) = x^2$ as x approaches 3 equals the value of $h(x)$ when $x = 3$." Functions possessing this property are called *continuous*. They are the functions whose graphs have no breaks or jumps.

(M) DEFINITION. We call the function f continuous at $x = a$ if and only if

$$\lim_{x \to a} f(x) = f(a).$$

Not all functions are continuous and it is not always easy to determine the value of L in (L). For example, if we let

$$g(x) = \frac{\sin x}{x}$$

then g is not defined at $x = 0$ and, hence, cannot be continuous at that point. $(g(0)$ does not exist because of the zero in the denominator.) Hence, it is uncertain whether or not

$$\lim_{x \to 0} g(x)$$

exists. And if it exists, its value is not obvious. It can be proved that

(N) $$\lim_{x \to 0} g(x) = \lim_{x \to 0} \frac{\sin x}{x} = 1$$

so that here, $L = 1$.

Incidentally, you must not conclude that the limit in (N) equals 1 because both the numerator and the denominator of the expression for $g(x)$ approach zero as x approaches zero. This same phenomena occurs in $(\sin \pi x)/x$ but

$$\lim_{x \to 0} \frac{\sin \pi x}{x} = \pi.$$

In fact, in the definition of $f'(x)$ given in equation (I), the difference quotient on the right-hand side always formally approaches an expression that "looks like zero divided by zero." But, as we have seen, $f'(x)$ represents the slope of the tangent to the graph of $y = f(x)$ and so can take on many different values.

Moreover, the expression $0/0$ has no meaning in mathematics. Part of the power and beauty of calculus comes from the easy manner with which it assigns meaning to expressions—like the difference quotient $D(\Delta x)$—which are *tending* toward the *undefined* object $0/0$.

The other type of limit—the one encountered in the definition of the integral—is defined in a manner analogous to definition (L):

(P) DEFINITION. We say that the limit as n approaches infinity of A_n is L and we write

$$\lim_{n \to \infty} A_n = L$$

if and only if for each $\varepsilon > 0$ there exists an integer $N > 0$ such that

$$n > N \text{ implies } |A_n - L| < \varepsilon.$$

Here, as in definition (L), ε is a measure of precision given in advance. We must then produce the positive integer N with the property that A_n will be closer to L than ε whenever n is larger than N. Notice that although the mysterious, self-swallowing symbol for infinity appears in the symbol

$$\lim_{n \to \infty} A_n = L$$

it does not appear in the working part of definition (P). The definition says nothing about "infinity." It says only that A_n must be "arbitrarily close" to L whenever n is sufficiently large." The ε measures the "arbitrary closeness" and the N provides the measure of sufficiently large.

Let's look at a simple example and formulate it in the nature of a theorem as we did for the example following definition (L).

THEOREM.

$$\lim_{n \to \infty} \left(\frac{6}{n} + 2 \right) = 2.$$

PROOF. Let $\varepsilon > 0$. Choose $N = \underline{\quad}$. Then $n > N$ implies

$$|(6/n + 2) - 2| = |6/n|$$
$$= 6/n$$
$$< 6/N.$$

Notice that this last expression will be *less* than ε provided the denominator exceeds $6/\varepsilon$, that is, if

$$6/\varepsilon < N.$$

Since we must choose an integer value for N, we will choose

$$N = \langle 6/\varepsilon \rangle$$

where the notation indicates

$$N = \text{"the first integer larger than } 6/\varepsilon.\text{"}$$

This is the value of N to be placed in the blank space and this choice completes the proof.

Recapitulating:

Let $\varepsilon > 0$. Choose $N = \langle 6/\varepsilon \rangle$. Then $n > N$ implies $n > 6/\varepsilon$. Hence,

$$\left| \left(\frac{6}{n} + 2 \right) - 2 \right| = \left| \frac{6}{n} \right|$$
$$= \frac{6}{n}$$
$$< \frac{6}{N}$$
$$\frac{6}{\dfrac{6}{\varepsilon}}$$
$$= \varepsilon.$$

In this example, incidently, it was easy to guess that $L = 2$ since the quantity $6/n$ obviously tends to zero as n becomes arbitrarily large so that $6/n + 2$ tends to 2. But, as in the case of discontinuous functions and definition (L), the value of L in definition (P) may be extremely difficult to determine. One such example that we have met before and that lies far beyond our grasp is

$$\lim_{n \to \infty}\left(1 + \frac{1}{4} + \frac{1}{9} + \frac{1}{16} + \ldots + \frac{1}{n^2}\right) = \frac{\pi^2}{6}.$$

I should point out that neither the *definition* of the limit nor the *proofs* of even the simplest theorems are ordinarily taught in freshman calculus. Conceptually, our work in this section is beyond freshman calculus. In these courses students mainly deal with limits of the form

$$\lim_{x \to a} f(x) = L$$

where *f* is continuous. They *evaluate* such limits by substituting the number *a* for *x* and writing $L = f(a)$.

So, you should not feel deficient if the definitions or the proofs have seemed difficult and not completely intelligible. But if you have been able to understand, even partially, these ideas, you will find college calculus far easier to comprehend than the pictures of Picasso or the poetry of Ezra Pound.

Moreover, when you go seriously to calculus, you find it equals in beauty both the pictures and the poetry. And you will find it far more powerful.

POWER AND BEAUTY

Phillip E. B. Jourdain wrote

> The extraordinary power of calculus in dealing with complicated questions lies in the fact that the question can be split into an infinity of simpler ones which can all be dealt with at once because of the wonderfully economical fashion in which calculus deals with variables.[5]

Exactly. And calculus possesses not just power but rather *extraordinary power*—a commanding force derived from simplicity and economy. In calculus, simplicity and economy yield power. And in calculus—as in all of mathematics—"simplicity" and "economy" are just other words for "beauty." The book of calculus is written in symbolic poetry. The poetry gives it power.

An excellent example of Jourdain's observation concerning the splitting of a complicated question "into an infinity of simpler ones" lies in the use of the integral to compute the area under the graph of $y = f(x)$. Here, we reduced the problem of finding the area of the whole region into a collection of simpler problems by decomposing the region into n "rectangular" strips. Then we let n tend to infinity and simultaneously obtained the area A and the integral of $f(x)$ from a to b. And we found

$$A = \int_a^b f(x)dx.$$

The integral, of course, has many interpretations other than area, just as the derivative has mathematical and physical interpretations other than as tangent lines or velocities. For example, if you are told by an actuary that your life expectancy follows a normal distribution with, say, mean equal to 68 years and standard deviation equal to 7 years, then he is telling you something about an integral. Precisely, you have been told that the probability that you will live longer than x years is

$$\frac{1}{7\sqrt{\pi}} \int_x^\infty e^{-\frac{1}{2}\frac{(x-68)^2}{49}} dx.$$

The fact that this integral (with an infinite upper limit) and the esoteric transcendental numbers π and e describe human lifetimes simultaneously reinforces the reality of the power of calculus and Wigner's point about the "unreasonable effectiveness" of mathematics.

Since the number e is less familiar than π, let's look at another example in which it occurs. We begin with a question from pure mathematics.

The derivative of a function f is another function f', defined by equation (*I*). It is natural to wonder about those functions that are equal to their own derivative, that is, those functions that remain *invariant* under the differentiation process. More precisely: let's find all functions f for which $f'(x) = f(x)$ for all values of x.

One way to proceed is to suppose that $f(x)$ has a power series representation for all real values of x. Let's assume

(1) $$f(x) = a_0 + a_1 x^1 + a_2 x^2 + \ldots + a_n x^n \ldots .$$

(The three dots in addition to the perioid on the right-hand side of (1) indicate that the series continues forever in this fashion. For this reason it is called an *infinite series*. This particular infinite series is called a *power series* because each term is a constant multiplied by a power of x. For our purposes it is sufficient to think of a power series as an "infinitely long polynomial."

It can be proved that if f can be represented by such a series then

(2) $$f'(x) = a_1 + 2a_2x + 3a_3x^2 + \ldots + na_nx^{n-1} + \ldots.$$

(Two "facts" are needed to prove (2): i) a power series can be differentiated "term by term," and ii) the derivative of $t(x) = ax^m$ is $t'(x) = max^{m-1}$.) Hence, $f(x) = f'(x)$ implies

(3) $$a_0 + a_1x + a_2x^2 + a_3x^3 + \ldots + a_{n-1}x^{n-1} + a_nx^n + \ldots$$
$$= a_1 + 2a_2x + 3a_3x^2 + 4a_4x^3 + \ldots + na_nx^{n-1} + \ldots.$$

Since $f(x) = f'(x)$ holds for all x, so does equation (3). It can be proved that this happens only when the series on each side of equation (3) are identical; that is, the coefficient of equal powers of x are identical. Therefore,

$$a_0 = a_1$$
$$a_1 = 2a_2$$
$$a_2 = 3a_3$$
$$a_3 = 4a_4$$

$$\cdots$$

$$a_{n-1} = na_n$$

$$\cdots$$

The equations yield, in particular

$$a_1 = a_0$$

(4) $$a_2 = \frac{a_1}{2} = \frac{a_0}{2 \cdot 1}$$

$$a_3 = \frac{a_2}{3} = \frac{a_0}{3 \cdot 2 \cdot 1}.$$

Continuing the analysis yields (as you would expect):

$$a_4 = \frac{a_0}{4 \cdot 3 \cdot 2 \cdot 1}$$

$$a_5 = \frac{a_0}{5 \cdot 4 \cdot 3 \cdot 2 \cdot 1}$$

. . .

$$a_n = \frac{a_0}{n(n-1)(n-2) \ldots 3 \cdot 2 \cdot 1}$$

. . . .

For a given positive integer m, the product of m and all the positive integers less than m is called "m factorial" and is denoted by $m!$. Hence,

$$2! = 2 \cdot 1$$

$$3! = 3 \cdot 2 \cdot 1$$

. . .

$$n! = n(n-1)(n-2) \ldots 2 \cdot 1.$$

With this notation, equation (4) may be written as:

$$a_1 = a_0$$

$$a_2 = \frac{a_0}{2!}$$

$$a_3 = \frac{a_0}{3!}$$

$$a_4 = \frac{a_0}{4!}$$

. . .

$$a_n = \frac{a_0}{n!}.$$

Hence, all the unknown coefficients of $f(x)$ may be expressed in terms of the single coefficient a_0. When these values are substituted into equation (1), we obtain

$$f(x) = a_0 + a_0 x + \frac{a_0}{2!}x^2 + \frac{a_0}{3!}x^3 \ldots + \frac{a_0}{n!}x^n + \ldots.$$

"Factoring out" the common term a_0 yields

(7) $$f(x) = a_0\left(1 + x + \frac{x^2}{2!} + \frac{x^3}{3!} + \ldots + \frac{x^n}{n!} + \ldots\right).$$

Let's denote the expression in parenthesis by $h(x)$, that is,

(8) $$h(x) = 1 + x + \frac{x^2}{2!} + \frac{x^3}{3!} + \ldots + \frac{x^n}{n!} + \ldots.$$

Moreover, let's denote the real number a_0 in equation (7) by the simpler symbol a. Thus, equation (7) becomes

(9) $$f(x) = ah(x).$$

Our analysis provides an outline of the proof of the following result, which may be proved rigorously:

THEOREM. If $f'(x) = f(x)$ holds for all x, then $f(x) = ah(x)$ where a is an arbitrary real number and $h(x)$ is given by equation (8).

Consequently, the only functions that remain invariant when they are differentiated are multiples of the function $h(x)$.

The function $h(x)$ is one of the most important and remarkable functions in all of mathematics. If we set $x = 1$ we obtain

(10) $$h(1) = 1 + 1 + \frac{1}{2!} + \frac{1}{3!} + \frac{1}{4!} + \ldots \frac{1}{n!} + \ldots.$$

The infinite series on the right-hand side of (10) is a series of constants and this series converges to the famous transcendental number e. Thus,

(11) $$e = 1 + 1 + \frac{1}{2!} + \frac{1}{3!} + \frac{1}{4!} + \ldots + \ldots.$$

(The notion of convergence of an infinite series is itself *defined* by a certain limit. For the infinite series (11) the definition says:

$$e = \lim_{n \to \infty}\left(1 + 1 + \frac{1}{2!} + \frac{1}{3!} + \ldots + \frac{1}{n!}\right).$$

This is the *meaning* of equation (11).)

It is not obvious that e is transcendental but it is nevertheless true. Being a transcendental real number, e is also irrational, which means that it has no finite or repeating decimal representation. Obviously, from (11), e is larger than 2 and it is not difficult to show that e is less than 3. Highly accurate rational approximations to e are known. A rational approximation of e to nine decimal places is 2.718281828.

It is not difficult to prove that for any *rational* number r, the value of $h(r)$ given by (8) is

$$h(r) = e^r.$$

That is, "$h(r)$ equals e raised to rth power," for any rational number r. Consequently, $h(x)$ is written, for all x, as

(12) $$h(x) = e^x.$$

If $f(x) = f'(x)$ holds for all x then

$$f(x) = ae^x$$

for some real number a.

Thus our problem in pure mathematics has as its solution the exponential function and its multiples.

Now let's consider a problem in applied mathematics. Suppose a "substance" (such as a colony of bacteria or an investment account) either grows or shrinks continuously with time at a rate proportional to the amount present. If y denotes the amount of substance present at time t, then the growth (or shrinking) rule is

(13) $$\frac{dy}{dt} = by$$

where b denotes the "constant of proportionality." Equation (13) then provides a mathematical model for the growing substance. This model fits many physical phenomena. For example, equation (13) could represent the growth of an account invested at a *continuously compounded interest rate* b. In this case, equation (13) says "the rate of growth of the account equals, at any time, the amount present multiplied by the interest rate."

Similarly, equation (13) may represent behavior of material shrinking because of radioactive decay. In this case, the proportionality constant b will be negative, indicating a negative rate of change of y with respect to t.

In order to solve equation (13), we must find a function $y = f(t)$ with the property that

$$\frac{dy}{dt} = f'(t) = by = bf(t)$$

that is

$$f'(t) = bf(t).$$

Except for the presence of the constant b, this is the problem in pure mathematics we have just considered. You will not be surprised to learn that the solution to equation (13) is the exponential function.

In fact, if we add the "initial condition" $y = c$ when $t = 0$—that is, the substance begins with amount c—the solution is

(14) $y = ce^{bt}.$

In the continuous interest example, equation (14) says:

> If c dollars are invested at an annual interest rate $100b$ %
> compounded continuously, then after t years the value of
> the account is $y = ce^{bt}$.

In this most practical of examples, you see the power and beauty of calculus made manifest. The power resides in the process that leads to the specific solution to the problem—a solution lying beyond the reach of anyone unfamiliar with calculus. The beauty comes from the simplicity and economy of this process and from the elegant form of the solution

$$y = ce^{bt}$$

and from this function's connection with the deeper notions of transcendental numbers and infinite series.

Another illustration of the extraordinary power of calculus lies in the ease with which it deals with bodies moving with constant acceleration. For illustration, let's consider a falling object—a stone, say, thrown straight down from the top of a tall building. A sketch appears in figure 67.

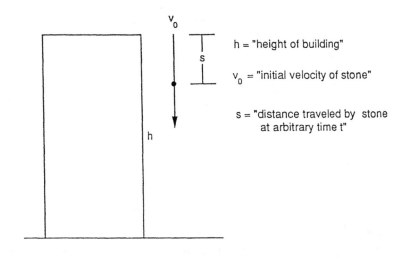

Figure 67: Falling stone

If we assume that the only force acting on the stone as it falls is the grav-itational force, then we know the stone's acceleration will be the acceleration due to gravity. This constant acceleration, traditionally denoted by g, is approximately 32 feet per second, per second, or 32 ft/sec^2. Thus, if we denote the velocity of the falling stone by v, we have

$$(15) \qquad \frac{dv}{dt} = g.$$

Since the derivative of velocity with respect to time equals acceleration (by definition), equation (15) states in mathematical terms the phrase: "the acceleration of the stone equals g." Remarkably, the equations of motion of the stone follow easily from (15) and the most elementary of calculus techniques.

Let's take the antiderivative of both sides of (15). On the left, "the" anti-derivative of $\frac{dv}{dt}$ is v and, on the right, "the" antiderivative of g is gt. Hence

$$(16) \qquad v = gt + c_1$$

where c_1 is a constant that appears because, as we have seen, antiderivatives are not unique but any two antiderivatives of a given function differ, at most, by a constant.

Let's suppose we know the initial velocity with which the stone was flung downward. Call this initial velocity v_0. Thus, the velocity of the stone at time $t = 0$ is v_0. Hence, $t = 0$ implies $v = v_0$. Substituting these values into equation (16) gives:

$$v_0 = g \cdot 0 + c_1$$

or

$$v_0 = c_1.$$

Equation (16) then becomes

(17) $$v = gt + v_0$$

an equation that provides the velocity of the stone at any time t.

If, as in figure 67, we let s denote the distance traveled by the stone at arbitrary time t, then, by definition of velocity,

$$v = \frac{ds}{dt}.$$

Therefore, equation (17) becomes

(18) $$\frac{ds}{dt} = gt + v_0.$$

If we take antiderivatives of both sides of equation (18), we obtain

(19) $$s = \frac{1}{2}gt^2 + v_0 t + c_2$$

where c_2 is another constant. But we know (see figure 67) that $s = 0$ when $t = 0$. Thus

$$0 = \frac{1}{2} \cdot g \cdot 0^2 + v_0 \cdot 0 + c_2$$

or

$$0 = c_2.$$

Equation (19) then becomes

(20) $$s = \frac{1}{2}gt^2 + v_0 t.$$

Equations (17) and (20) completely describe the motion of the falling stone—the former equation specifying the stone's velocity at any time and the latter equation giving its position. Additional information can be obtained from these equations by simple manipulation.

For example, suppose a stone is dropped off a 100-foot-tall building. With what velocity does the stone strike the ground? In this case, $v_0 = 0$ because the stone is dropped rather than being flung downward. Equations (17) and (20) then become

$$v = gt$$

and

$$s = \frac{1}{2}gt^2.$$

The stone strikes the ground when $s = h$, the height of the building. Substituting this into the second equation we have

$$h = \frac{1}{2}gt^2$$

or

$$t^2 = \frac{2h}{g}.$$

This gives

$$t = \sqrt{\frac{2h}{g}}$$

at the time at which the stone strikes the ground. Substituting this value of t into the equation for v gives

$$v = \left(\sqrt{\frac{2h}{g}}\right)g$$

as the velocity with which the stone strikes the ground. This can be simplified to

$$v = \left(\sqrt{\frac{2h}{g}}\right)g$$

$$= \sqrt{g}\sqrt{g}\sqrt{\frac{2h}{g}}$$

$$= \sqrt{g}\sqrt{2h}$$

$$= \sqrt{2gh}.$$

Hence, the stone strikes the ground with velocity

$$v = \sqrt{2gh}.$$

In our example $h = 100$ ft. Substituting this value into the preceding equation and remembering that $g = 32$ ft/sec^2 gives

$$v = \sqrt{2\left(32\text{ft} / \sec^2\right)\left(100\text{ft}\right)}$$
$$= \sqrt{6,400\text{ft}} / \sec$$
$$= 80\text{ft} / \sec.$$

Thus, the "striking velocity" is 80 ft/sec.

Galileo observed and recorded the behavior of falling bodies. Newton, who was born the year Galileo died, wanted to *understand* this motion. And *understanding* required a mathematical description of motion. The great man struggled to develop machinery sufficient to produce a description. The machine he built is called calculus. With it, college freshmen can reproduce equations (17) and (20) with cool aplomb and almost frightening ease.

As a final illustration of the power of calculus, let's consider the question of determining the *extreme values* of a given function. (An *extreme value* is either a maximum value or a minimum value.) Suppose $y = f(x)$ is defined for all real numbers x with $a \leq x \leq b$. What are the maximum and minimum values of $f(x)$ for $a \leq x \leq b$?

The set of real numbers x with $a \leq x \leq b$ is called a *closed interval* and is denoted by $[a, b]$. In this language the problem becomes

> Let f be defined on $[a, b]$. What are the maximum and minimum values of $f(x)$ on $[a, b]$?

With this generality, the question has no answer since there exist functions and closed intervals such that there is neither a maximum nor a minimum function value. But for "nice" functions the question has, at least, a theoretical answer.

You will recall that (intuitively) a continuous function is one whose graph is unbroken. It seems clear—again on intuitive grounds—that such a function would attain a largest and a smallest value on a closed interval $[a, b]$. This intuitive result can be rigorously proved and the resulting theorem says:

If f is continuous on $[a, b]$ then $f(x)$ attains a maximum and a minimum value on $[a, b]$.

The proof of this theorem requires a deeper knowledge of the properties of real numbers than calculus students ordinarily possess and it belongs to a more advanced course. You should notice that the theorem only asserts the *existence* of the extreme value of $f(x)$. It does not tell you how to find these values. Things are simpler if the function is differentiable.

Recall that differentiable functions are those functions whose graphs are sufficiently smooth so that, at each point, they possess a tangent line. The graph of such a function appears in figure 68. Here, the graph has local high or low points between a and b at x_1, x_2, x_3, and x_4. At each of these local extreme points, the tangent to the graph is horizontal, that is, the tangent has zero slope. Consequently, $f'(x_1) = f'(x_2) = f'(x_3) = f'(x_4) = 0$.

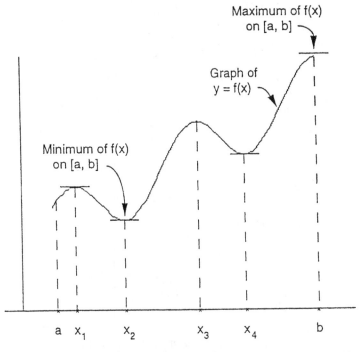

Figure 68: Maxima and minima

It is geometrically evident that this must always be the case. If the graph of $y = f(x)$ has a "local high point" at some w where $a < w < b$, then the graph must have a horizontal tangent line at the point $(w, f(w))$. Hence, $f'(w) = 0$. But any maximum value of $f(x)$ for $a < x < b$ must certainly be a local high point of the graph of $y = f(x)$. Consequently, the only candidates for maximum values of $f(x)$ between a and b occur at values of x for which $f'(x) = 0$. Similarly, the only possibilities for minimum values of f occur at points where $f'(x) = 0$.

However, not all values of x between a and b for which $f'(x) = 0$ will be values where $f(x)$ attains a local maximum or a local minimum. To see this, consider the function $f(x) = x^3$ and the closed interval $[-1, 1]$. Here, $f'(x) = 3x^2$ so that $f'(0) = 0$. But the graph of $y = x^3$ does not have a local maximum or a local minimum when $x = 0$ (see figure 69). The best that can be said (and this is saying a lot) is: If $f(x)$ has a maximum or a minimum at w strictly *between* a and b, then $f'(w) = 0$.

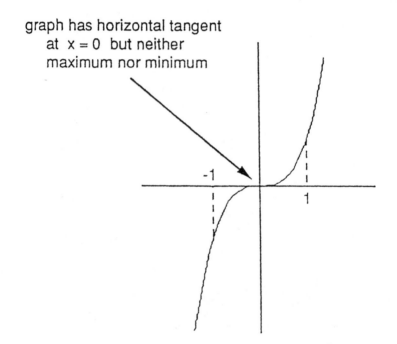

graph has horizontal tangent
at x = 0 but neither
maximum nor minimum

-1

1

Figure 69: Horizontal tangent line

The maximum, or the minimum, value of $f(x)$ on $[a, b]$ could occur at one of the endpoints. In figure 68, for example, the maximum occurs at $x = b$ while the minimum appears at the interior point x_2. But for all these qualifications, the derivative provides a powerful tool for the location of extreme values of a function:

The extreme value of a differentiable function f on $[a, b]$ occurs at $x = a$ or $x = b$ or values of x *between* a and b where $f'(x) = 0$.

Thus, we can determine the maximum and minimum values of a differentiable function by simply making a list. We first compute the derivative, $f'(x)$, of $f(x)$. Then we solve the equation

$$f'(x) = 0$$

for x. Suppose the solutions that lie *between* a and b are $x_1, x_2, x_3, \ldots, x_n$. Then we compute the values of $f(x)$ at these points and at a and b and we set down the list:

$$f(a), f(x_1), f(x_2), \ldots, f(x_n), f(b).$$

The largest number in this list is the maximum value of $f(x)$ on $[a, b]$ and the smallest number is the minimum value.

There are more efficient methods for determining extreme values but none are as simple or as intellectually economical as this. And few ideas in all of mathematics have greater practical value. The method allows you—in an elementary prescriptive manner—to determine the extreme values of a differentiable function and, therefore, to determine the maximum and minimum values of any real-world quantity that can be expressed in terms of such a function.

Let's illustrate the method by applying it to an extremely practical problem—that of designing a tin can to hold a given volume, which is to be made of the smallest amount of metal. When we are finished, we will make an observation that suggests even such a seemingly dull and routine problem may have, underneath it, an aesthetic component.

An ordinary tin can is a right circular cylinder, closed at both ends. If we let h denote the height of the cylinder and x denotes the radius of the base, then the volume, V, is given by

(21) $$V = \pi x^2 h.$$

The surface area of the cylinder can be determined by adding the areas of the cylinder's top and bottom to its lateral area. The top and bottom are each circles of area πx^2. The lateral area, when unrolled and laid flat, equals the area of a rectangle whose length is the circumference of the cylinder's top (or bottom) and whose height is the height, h, of the cylinder. Total area then equals $2\pi x^2 + 2\pi xh$. So, if we let S denote the surface area of the cylinder, we have

(22) $$S = 2\pi x^2 + 2\pi xh.$$

The cylinder and the rolled-out surface area are shown in figure 70.

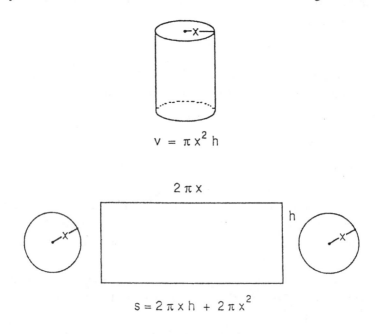

Figure 70: Volume and area of a cylinder

Now, cylinders that have the same volume can have various heights. You can make, for example, a tall, thin cylinder that will hold exactly one quart of liquid or you can make a short, fat cylinder containing the same volume. And, in between these extremes, you can design an infinity of one-quart tin cans. You expect these cans to require varying amounts of metal for their construction.

Our problem is to determine the relative dimensions of the right circular cylinder with *fixed volume* given by equation (21) such that the surface area given by (22) has minimum value.

As things stand, S is a function of two variables, x and h. But we can solve equation (21) for h and thereby determine h in terms of x alone. Then we can apply our method to the new function S.

Solving equation (21) for h gives:

$$h = V/\pi x^2.$$

Substituting this value into equation (22) gives:

(23)
$$S = \frac{2V}{x} + 2\pi x^2.$$

Since our cylinder may theoretically have any positive radius, the function S is defined for all $x > 0$ and our problem is to determine the minimum value of S, if it exists, on this set rather than on some closed interval. But, in order to use our method, we select a value of a and b where $a > 0$, $b > 0$, a is "close" to zero, and b is "very large." We'll first determine the minimum value of S on $[a, b]$.

The derivative of x^2, as we have seen, is $2x$. It can be shown that the derivative of $1/x$ is $(-1/x^2)$. Thus

$$S'(x) = -\frac{2V}{x^2} + 4\pi x.$$

Hence,

$$S'(x) = 0$$

implies

$$-\frac{2V}{x^2} + 4\pi x = 0.$$

Solving this equation for x gives

(24)
$$x = \sqrt[3]{\frac{V}{2\pi}}$$

that is, x equals the cube root of $V/2\pi$. If S has a minimum value on $[a, b]$, it must occur at this value of x or at the endpoints a or b. Substituting the x value into equation (23) gives

(25)
$$S = \frac{2V}{\sqrt[3]{\frac{V}{2\pi}}} + 2\pi \left(\sqrt[3]{\frac{V}{2\pi}} \right)^2.$$

For large values of x, $\dfrac{2V}{x}$ is close to zero, and equation (23) shows that S then behaves like $2\pi x^2$. Similarly, equation (23) shows that for x near zero, S behaves like $\dfrac{2V}{x}$. Thus, S is very large for x large or for x near zero. Hence, of the three numbers in our list—$S(a)$, $S(b)$, and the value in (25)—the latter value is certain to be the smallest.

Thus, the value given by equation (25) is the minimum value of S on $[a, b]$. And, since S increases as does $\dfrac{1}{x}$ to the left of a and as does x^2 to the right of b, this value is the minimum value of S for all $x > 0$.

We know from (21) that

$$h = \frac{V}{\pi x^2}.$$

Dividing by x gives

(26)
$$\frac{h}{x} = \frac{V}{\pi x^3}.$$

Substituting the value of x from (24), which gives S the minimum value, into the right-hand side of (26) gives

$$\frac{h}{x} = \frac{V}{\pi \left(\dfrac{V}{2\pi} \right)}$$

or

$$\frac{h}{x} = 2.$$

Consequently, of all possible right circular cylinders having height h and radius x and fixed volume V, the cylinder with minimum surface areas is the one with

$$h = 2x.$$

That is, the cylinder of fixed volume having minimal surface area is the cylinder that has its height equal to the diameter of its base.

This seems to me an aesthetically pleasing result. When you look at a kitchen shelf of tin cans at eye level, you can see a row of rectangles. If these cans each contain the same volume, then the one that solves our maximum-minimum problem will appear on the shelf as a square. A typical row of such rectangles appears in figure 71. The shaded rectangle in the figure is a square.

Figure 71: Side view of cylinders

In 1876, Gustav Fechner attempted to measure aesthetic preferences by having subjects examine a series of ten rectangles of various length-width ratios and select the one they found most beautiful. Fechner found a preference for the ratio known as the "golden section, which is approximately the ratio of 1 to 1.61803." I. C. McManus says of Fechner's work:

> The experiment has been enormously influential, being accepted by many non-scientists, and indeed by many psychologists, as incontrovertible scientific proof of the superiority of the golden section.[7]

However, when McManus himself performed similar experiments, he found two clear aesthetic preferences, one based on the golden rectangle and the other based on a square.

In the row of tin cans shown in figure 71, the third from the left is a square. None of the rectangles in the figure have golden section length-width ratios. It is interesting and pleasing that the cylinder with fixed volume and minimum surface, found by the methods of calculus, also represents a confirmed aesthetic choice.

Chapter 9

PATTERNS AND PARADOX

When he came to the end of his mathematical life, G. H. Hardy (1877–1947) wrote a hauntingly beautiful book called *A Mathematician's Apology*. In it he says:

> A mathematician, like a painter or a poet, is a maker of patterns. If his patterns are more permanent than theirs, it is because they are made with ideas.[1]

Yes. Mr. Hardy provides yet another metaphor for our subject: mathematics is a collection of *permanent patterns made of ideas.*

Hardy had no doubt of the permanence of mathematics: "Greek mathematics is permanent, more permanent even than Greek literature."[2] Others have argued for the permanence of art. Hemingway, for example, says brutally:

> A country finally erodes and the dust blows away, the people die and none of them were of any importance permanently except those who practiced the arts.[3]

Hardy, the mathematician, and Hemingway, the writer, would be in agreement had Hemingway considered mathematics an art. But I doubt Hemingway considered mathematics at all.

We've known from the outset of our study that mathematics is composed of ideas. And we have already seen many patterns. We saw how the preschool counting method of pointing and saying extended naturally to the formal notion of one-to-one correspondence. Thus we were led to the concept of counting sets as large as natural numbers themselves. The counting pattern is the same, whether we count jellybeans or infinite sets.

We looked at the binary operation of addition on the integers and noticed that its properties could be abstracted and placed down on more general sets in such a way as to preserve the pattern of operations. This gave us the fundamental algebraic object known as a group.

Had we pursued probability theory further we would have seen how much of this subject, and its many applications, fall into a few fundamental patterns. For example, many problems concerning game theory, vaccine testing, or random sampling can all be reduced to determining the outcome of an experiment involving the tossing of biased coins. Somewhere a physician wants to solve a genetics problem and a biologist faces a contagion question involving laboratory animals. Simultaneously, a physicist deals with a problem of heat exchange between isolated bodies. Evidently, three different problems in three areas of application—until along comes a mathematician who points out the common pattern. All three require for solution only the probabilistic determination of all manners in which certain colored balls can be selected from appropriate boxes.

Literal—and often wonderful—patterns occur routinely in counting problems. We saw an example earlier when we noted the consistency of the method of counting the number of combinations of n objects taken r at a time. The answer, in every case, turns out to be

$$C(n, r) = \binom{n}{r}.$$

Another such example, one that involves the concept of one-to-one correspondence, follows:

Consider the question "In how many ways can eight oranges be distributed to the three children?" (It is natural to assume that the oranges are not distinguishable from one another but the children are.) The problem can be solved in the dull and uninteresting manner of simply listing all possibilities $(8, 0, 0)$, $(7, 1, 0)$, $(7, 0, 1)$, . . . , $(0, 0, 8)$, where each triple indicates the number of oranges given to each child, respectively.

The trouble with this method, besides its tedium, is that it does not help us to solve the problem if the number of oranges or the number of children is changed. It does not provide a pattern.

Consider the symbol

$$-- x --- x ---$$

and interpret it to mean that the number of spaces before the first x is the number of oranges given to the first child, the number of spaces between the x's is the number of oranges for the second child, and the number of spaces after the second x represents the number of oranges given to the third child. So, the above symbol corresponds to the distribution (2, 3, 3) while

$$- - - - - - - x - x$$

corresponds to (7, 1, 0). Clearly there exists a one-to-one correspondence between the collection of symbols of this type and the set of possible distributions of the oranges. Thus there are the same number of distributions of oranges as there are symbols of this type. But this is exactly $\binom{10}{2} = 45$, the number of ways to choose the spaces for the two x's.

And the pattern is clear. If there are r oranges and c children, then the number of distributions is $\binom{r + c - 1}{c - 1}$.

Moreover, the concept of pattern *defines* an entire branch of mathematics called *topology*. Topologists study those properties of mathematical objects that remain unchanged under certain kinds of transformations. To a topologist a circle and a triangle represent the same kind of closed curve because one can be deformed into the other by means of a continuous deformation. You may have to stretch and smooth the triangle to turn it into the circle, but you won't have to tear it. Similarly, a coffee cup made of malleable clay can be shaped, without tearing, into the form of a doughnut. You just fatten up the handle by squeezing the other material into it. A topologist, then, is a mathematician who does not know the difference between a doughnut and a coffee cup.

A transformation between mathematical objects that preserves certain operations on the objects is called an *isomorphism*. Two objects are isomorphic if they are the same except for appearance—the pattern of the operations in either setting is identical with that of the other. Much of mathematics deals with the concept of isomorphism, a notion that exemplifies the idea of pattern.

FALLACY

Interwoven with the patterns of mathematics are puzzlements called *paradoxes* and flat-out errors known as *fallacies*. The ordinary meaning of "paradox" is that of a "correctly deduced self-contradictory statement." We saw an example of such a paradox in Russell's paradox. If we allow the existence of a set that simultaneously is, and is not, an element of itself, the paradox is inevitable. (Or a barber who both does, and does not, shave himself.) The way out, in this case, is to forbid any set from ever being an element of itself.

Often, however, mathematicians use the term "paradox" to refer to a proposition that can be proved but that has a conclusion egregiously out of order with intuition or experience. We shall interpret "paradox" in this manner.

On the other hand, the term "fallacy" denotes a result that is incorrect but which, at first glance, seems to have been correctly deduced. Thus a fallacy is something more than a simple mistake. A fallacy results from an error of reasoning that occurs with sufficient subtlety that it merits examination. (Of course, "subtle" is a relative term. What seems subtle to a novice may seem blatant to an experienced mathematician.)

We encountered a fallacy earlier with the little argument purporting to show that $2 = 1$. The not-so-subtle error here came in the line where both sides of an equation were divided by $(x - y)$, which amounts to dividing by zero since, in this demonstration, $x = y$. Let's now look at some other fallacies before we turn to paradoxes. Please remember that the conclusion of each fallacy is INCORRECT.

Fallacy 1. *Another "proof" that $2 = 1$.*

Solve the equations

$$\frac{x-5}{x-1} = \frac{x-5}{x-2}.$$

Solution. Since

$$\frac{x-5}{x-1} = \frac{x-5}{x-2}$$

and the fractions have identical numerators, then their denominators must be identical. Thus,

$$x - 1 = x - 2.$$

Subtracting x from both sides gives:

$$-1 = -2.$$

Hence,

$$1 = 2.$$

The mistake comes in the first step:

$$\frac{a}{b} = \frac{a}{c} \Rightarrow b = c \text{ or } a = 0.$$

A correct solution is

$$\frac{x-5}{x-1} = \frac{x-5}{x-2} \Leftrightarrow (x-5)(x-2) = (x-1)(x-5)$$

$$\Leftrightarrow x^2 - 7x + 10 = x^2 - 6x + 5$$

$$\Leftrightarrow 7x - 10 = 6x - 5$$

$$\Leftrightarrow x = 5.$$

So, the solution is $x = 5$. Notice that when $x = 5$, the original equation becomes

$$\frac{0}{4} = \frac{0}{3}$$

which is correct.

FALLACY 2. $\sqrt{2} = 2$.

ARGUMENT. Consider a square with sides of length 1. Construct a square of "stair step lines"

$$s_1, s_2, s_3, \ldots, s_n, \ldots$$

as shown in figure 72. The diagonal of the square is the dotted line d. The stair-steplines are marked with arrows. Let $\ell(d)$, $\ell(s_1)$, $\ell(s_2)$, ..., $\ell(s_n)$, ... denote the lengths of these "lines." Clearly s_n becomes arbitrarily close to the diagonal d as n becomes arbitrarily large. Thus $\ell(s_n)$ becomes arbitrarily close to $\ell(d)$ as n grows without bound. Hence,

(1) $$\lim_{n \to \infty} \ell(s_n) = \ell(d).$$

But $\ell(s_n) = 2$ for each $n = 1, 2, \ldots$, since each stair step line has horizontal components with total length 1 and vertical components whose total length equals 1. Then (1) can hold only if $\ell(d) = 2$.

But the Pythagorean theorem says $\ell(d) = \sqrt{2}$. Hence, $2 = \sqrt{2}$.

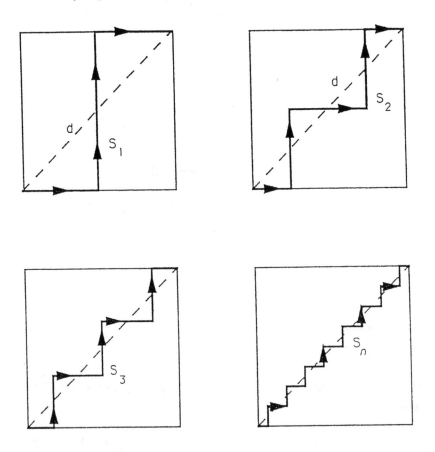

Figure 72: Approximation to the diagonal of a square

FALLACY 3. Every triangle is isosceles.

ARGUMENT. The argument uses high school plane geometry.

Consider the arbitrary triangle with vertices *a*, *b*, *c* of figure 73. Bisect the angle at vertex *a* and construct the perpendicular bisector of side *bc*. Let *p* denote the point of intersection of these two lines.

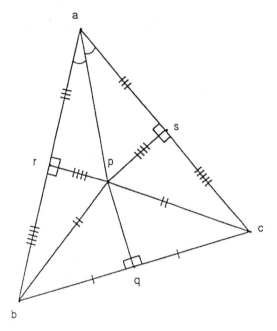

Figure 73: Fallacious argument that each triangle is isosceles

From *p* draw lines perpendicular to *ab* and *ac*. Let *r* and *s* be the respective intersection points. Construct lines *pb* and *pc*. Triangle *pqc* is congruent to triangle *pqb*. (Two sides and an included angle of one triangle equal two sides and an included angle of the other.) Thus lines *pb* and *pc* are the same length. Triangles *apr* and *aps* are congruent. (They have two equal angles by construction so their third angles must agree. But they also have the included side *ap* in common. So they are congruent.)

Hence lines *as* and *ar* have equal lengths and lines *pr* and *ps* have equal lengths. Then triangles *prb* and *psc* are congruent. (They are right triangles with two equal sides. So, the third sides are equal by the Pythagorean theorem.) Thus lines *br* and *cs* have equal lengths.

It follows that lines *ab* and *ac* are the same length. (The length of line *ab* equals the length of *ar* plus the length of *rb*. The length of *ac* equals the length of *as* plus the length of *sc*.)

Therefore, triangle *abc* is isosceles.

(The mistake occurs in the original construction. The point p falls within the triangle only when it is isosceles.)

FALLACY 4. $1 = 0$.

ARGUMENT. Let s be the sum of the infinite series $1 - 1 + 1 - 1 + \ldots$ So,

$$s = 1 - 1 + 1 - 1 + 1 - 1 + \ldots$$

Hence

$$s = (1 - 1) + (1 - 1) + (1 - 1) + \ldots$$
$$= 0 + 0 + 0 + 0 + \ldots$$

Therefore,

$$s = 0.$$

But also,

$$s = 1 - (1 - 1) - (1 - 1) - (1 - 1) - \ldots$$
$$= 1 - 0 - 0 - 0 - 0 \ldots$$
$$= 1 - 0$$
$$= 1.$$

Hence,

$$1 = 0.$$

(The argument fails at the first step. Only under certain conditions can parentheses be inserted into an infinite series. These conditions do not hold for the series $s = 1 - 1 + 1 - 1 + \ldots$.)

Were fallacy 4 correct, it could be written in the form

(2) $$1 = 0 + 0 + 0 + 0 + \ldots$$

which would have interesting philosophical interpretations. If we think of 1 as somehow denoting the "oneness of the universe" and 0 as representing "nothing," then (2) asserts that you can make the complete universe out of "nothings" as long as you have infinitely many of them.

(Equation (2) can assert whatever it wishes and we will have no quarrel. For equation (2) has nothing to do with mathematics, which is the subject under consideration. Equation (2) is simply a set of meaningless marks written on paper.)

However, mathematics does contain *many correct* results of such a counterintuitive nature that at first glance may seem as meaningless. As we know, a result of this type is called a paradox.

PARADOX

The nature of probability theory makes it a rich source of paradoxes. We will examine six of them.

(a) *The Paradox of the Second Child*

> A couple has two children. If at least one of the children is a girl, what is the probability that both are girls?

SOLUTION. Since births of girls and births of boys occur with almost equal frequency, we begin by *assuming* that they occur with equal probability. Thus the probability of a girl at each birth is $\frac{1}{2}$ and the same is true for the probability of a boy. We also *assume* that outcomes of each birth are independent of one another. Consequently—since our couple has two children—we may think of the problem situation as being equivalent to an experiment consisting of two independent tosses of a fair coin. The sides of the coin are marked g for girl and b for boy.

At this stage the intuitive answer to the problem is that the probability of two girls must be $\frac{1}{2}$ since the coin lands either g or b with probability $\frac{1}{2}$. To see that this intuitive answer is incorrect, consider the sample space

$$S = \{(g, g), (g, b), (b, g), (b, b)\}.$$

Here, for example, the element (b, g) means "first child boy and second child girl." The other elements of S have analogous interpretations. Thus S represents all possible outcomes of the two-children "experiment."

Since b or g occurs at each stage with probability $\frac{1}{2}$, the coin is fair and the equiprobable probability measure applies to S. Let A and B be the events:

$$A = \text{"both children girls"}$$

$$B = \text{"at least one child is a girl."}$$

Then

$$A = \{(g, g)\}$$

$$B = \{(g, g), (g, b), (b, g)\}.$$

Thus

$$A \cap B = \{(g, g)\}.$$

By definition

$$P(A \mid B) = \frac{P(A \cap B)}{P(B)}.$$

But

$$P(A \cap B) = \frac{n(A \cap B)}{n(S)} = \frac{1}{4}$$

and

$$P(B) = \frac{n(B)}{n(S)} = \frac{3}{4}.$$

Hence,

$$P(A \mid B) = \frac{\frac{1}{4}}{\frac{3}{4}} = \frac{1}{3}.$$

Therefore,

$$P(\text{"both children girls"} \mid \text{"at least one girl"}) = \frac{1}{3}.$$

A quick way to obtain the result of paradox (a) is to notice that, if we are given the information that at least one child is a girl, then the sample space reduces to

$$T = \{(g, g)(g, b), (b, g)\},$$

each element of which is as equally likely as any other. Only one of these three gives two girls. Thus its probability is $\frac{1}{3}$.

The nonintuitive aspect of this result usually stems from a misinterpretation of the given information. We are told only that "at least one child is a girl." We are *not* told which of the children is a girl. If we know, for example, that the oldest child is a girl, then the sample space reduces to

$$U = \{(g, g), (g, b)\}.$$

Thus the probability of both children being girls becomes $\frac{1}{2}$.

A more complicated version of the reasoning involved in this paradox received national attention as a result of controversy over a puzzle published in *Parade* magazine:[4]

(b) *The Birthday Paradox*

Twenty-three people are selected at random. What is the probability that at least two of them were born on the same day of the year?

SOLUTION. We *assume* that a year consists of 365 days and that each of these is equally likely as a birthday in the sense that a random selection of 23 people is equivalent to a random selection of 23 birthdays. Moreover, we assume the birthdays are independent of one another. Intuitively, it seems clear that the requested probability must be quite small. We have only 23 people and 365 possible birthdays. Few people, lots of possible birthdays.

We can just as easily solve the problem for a random selection of r people. Thus let r people be selected at random. Let $A(r)$ and $B(r)$ be the events:

$$A_r = \text{"no two of the people have the same birthday"}$$

and

$$B_r = \text{"at least two of the people have the same birthday."}$$

We think of the days of the year as bearing the numbers $1, 2, \ldots, 365$. Then January 1 becomes day number 1 and December 31 is day 365. To each of the r people in our random sample we assign the unique number from the list $1, 2, 3, \ldots, 365$, which constitutes that person's birthday. Then our sample space becomes:

$$S = \{(x_1, x_2, x_3, \ldots, x_r)\}$$

where each x_k is one of the numbers 1, 2, 3, . . . , 365. Since birthdays are assumed to be equally likely, we use the equiprobable measure P on S. We want to find $P(A_r)$. Since A_r and B_r are complementary events, corollary 1 of chapter 7 ensures

$$P(A_r) + P(B_r) = 1.$$

Thus it is sufficient to find $P(B_r)$. We know

$$P(B_r) = \frac{n(B_r)}{n(S)}.$$

According to the fundamental counting principle, a typical element of S may be chosen in

$$(365)(365) \ldots (365) = (365)^r$$

ways since there are 365 choices for each x_k in $(x_1, x_2, \ldots, x_r)$. Thus

$$n(S) = (365)^r.$$

A typical element of B_r looks like $(y_1, y_2, \ldots, y_r)$, where no two of y_k are identical. Thus y_1 may be chosen in 365 ways, y_2 in 364 ways, y_3 in 363 ways, and so on. Hence,

$$n(B_r) = (365)(364)(363) \ldots (365 - r + 1).$$

Therefore,

$$P(B_r) = \frac{(365)(364) \ldots (365 - r + 1)}{(365)^r}.$$

Figure 74 provides a table of values $P(A_r)$ for certain values r. Notice that when $r = 23$

$$P(A_r) = P(A_{23}) = .507.$$

Then the probability that at least two of a collection of twenty-three people have conflicting birthdays exceeds $\frac{1}{2}$. It's a more likely event than is getting heads on a single toss of a fair coin.

Once I knew a mathematics graduate student who partially supported himself by making bets related to this paradox. Mostly he bet with nontechnical people who would naively wager that a group of fifty people would not contain at least two with the same birthday. As you see from figure 74, his takers had no chance since $P(A_{50}) = .970$.

A_r = "at least two of r people have the same birthday".

r	$P(A_r)$
10	.117
20	.411
22	.476
23	.507
40	.891
50	.970
60	.994

Figure 74: The birthday paradox

My friend did quite well until one day he accepted a wager himself. A student of English bet him he could not memorize a randomly selected page of *Finnegans Wake*. "I not only failed to memorize the page I chose," my friend told me, "I couldn't even read it."

(c) *The Great Expectations Paradox*

> Players A and B toss a fair coin until it lands heads. A agrees to pay B one dollar if the coin lands heads on the first toss; two dollars if the first heads appears on the second toss; four dollars if on the third; eight dollars if on the fourth, and so on. Thus A pays B 2^{n-1} dollars if the first heads appears on the nth toss. How much should B be willing to pay A for the privilege of playing this game?

SOLUTION. Here we must deal with the countably infinite sample space

$$S = \{H, TH, TTH, TTTH, \ldots \}.$$

If we *assume* that the tosses are independent of one another then

$$P(H) = \frac{1}{2}$$

$$P(TH) = \frac{1}{2} \cdot \frac{1}{2} = \frac{1}{2^2}$$

$$P(TTTH) = \frac{1}{2} \cdot \frac{1}{2} \cdot \frac{1}{2} \cdot \frac{1}{2} = \frac{1}{2^4}$$

and so on. Let f denote the random variable that represents the possible winnings of player B. The range of f is

$$R(f) = \{1, 2, 2^2, 2^3, \ldots\}.$$

Also,

$$P(f = 1) = P(H) = \frac{1}{2}$$

$$P(f = 2) = P(TH) = \frac{1}{2^2}$$

$$P(f = 3) = P(TTH) = \frac{1}{2^3}.$$

Thus the expected value of f is

$$E(f) = \sum_{y \in R(f)} yP(f = y)$$

$$= 1 \cdot \frac{1}{2} + 2 \cdot \frac{1}{2^2} + 2^2 \cdot \frac{1}{2^3} + \ldots .$$

So

(3) $$E(f) = \frac{1}{2} + \frac{1}{2} + \frac{1}{2} + \ldots .$$

The right-hand side of (3) is a divergent infinite series. The nth partial sum of this series is (by definition)

$$S_n = \frac{1}{2} + \frac{1}{2} + \ldots \frac{1}{2}$$

where the sum contains n terms. So,

$$S_n = \frac{n}{2}.$$

Thus S, becomes arbitrarily large as n increasing without bound, that is

$$\lim_{n \to \infty} S_n = \infty.$$

But this means that $E(f)$ exceeds any given positive number. (We say "the expected value" of f is infinite.) Therefore, player B can expect to win an amount that exceeds any given number. Thus he should be willing to pay any given amount requested by A for the privilege of playing the game.

This paradox also bears the name "St. Petersburg paradox" because a version was first published in a journal of the St. Petersburg Academy in Russia.

(d) *Paradox of the Three Prisoners*

Three prisoners A, B, and C learn that two of them are to be pardoned and the choice has already been made by random selection. Prisoner A, who knows both elementary probability and the warden, reasons as follows: "The possibilities are that (A, B), (A, C), or (B, C) will be released. I appear in two of these. Since each choice is equally likely, the probability that I will be released is $\frac{2}{3}$. While it would not be fair to ask the warden if I am to be freed, he might give me the name of another prisoner who is to be released. I'll ask. But wait. Suppose I ask and he tells me, for example, that B is to be released. This eliminates the possibility (A, C) and the only remaining possibilities are (A, B) and (B, C). So my probability of being freed is reduced from $\frac{2}{3}$ to $\frac{1}{2}$. I won't ask. It's better not to know."

This paradox has been discussed by Martin Gardner.[5] It brought, he wrote "a flood of mail."

The paradox can be resolved by the construction of an appropriate sample space that takes into consideration what the warden might say and with what probability he might say it. The details are left as an exercise.

(e) *The Pairwise-Best Is Worst Paradox*

Consider three spinners as shown in figure 75. Spinner A always comes up 3. Because of the sizes of the sectors, spinner B gives values 2, 4, 6 with probabilities .56, .22, and .22, respectively. Spinner C points to 1 with probability .51 and to 5 with probability .49. Suppose two people select spinners and play. The highest number wins. It is obvious that A beats B with probability .56. (A always points to 3 and beats B if and only if B comes up 2.) Similarly, A beats C with probability .51. It is easy to see that B beats C with probability .62.

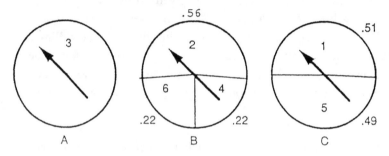

Figure 75: Three spinners for the pairwise-best is worst paradox

Therefore, if there are two players, A is best and C is worst.

Now suppose there are three players and, again, the highest number wins. It is routine to compute the probabilities: $P(\text{``}A\text{ wins''}) = .29$, $P(\text{``}B\text{ wins''}) = .33$, and $P(\text{``}C\text{ wins''}) = .38$. Now the results are reversed. (See figure 76.) Now A is worst and C is best.

This is the pairwise-worst-best paradox of Cohn R. Blyth.[6] It must say something profound about choice and preference in presidential elections. And it has other applications.

Consider the spinners as representing the effectiveness of antitoxins and suppose you are bitten by the swamp adder, "the deadliest snake in India."[7] They rush you to a jungle hospital. As your life fades, a physician prepares an injection.

"You are most fortunate," the physician tells you. "There are only three antitoxins available for swamp adder poison. And I have a vial of each. You see them there on the shelf labeled A, B, and C. Each has been thoroughly tested and rated on an effectiveness scale of 1 through 6. Antitoxin A has rating 3 for 100% of the population, while *B* has effectiveness 2, 4, and 6 for 56%, 22%, and 22% of the population, respectively. Substance *C* is rated 1 for 51% of the population and 5 for 49%."

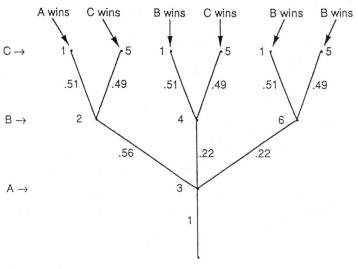

P("A wins") = (.56) (.51) = .29

P("B wins") = (.22) (.51) + (.22) (.51) + (.22) (.49) = .33

P("Cwins") = (.56) (.49) + (.22) (.49) = .38

Figure 76: Pairwise-best is worst paradox (three players)

Your eyesight is dim now, your body is numb. Faintly, you see him take down a vial. "The statisticians have determined," he says, "that *C* is best and *A* is worst. I will inject you with antitoxin *C*."

There is the sound of breaking glass. The physician speaks again.

"Unfortunately, vial *B* has fallen off the shelf and broken. Now I have only antitoxins *A* and *C*. Under these conditions *A* is best. Forget what I said before. I will give you an injection of *A*."

At this point the snake poison in your bloodstream is irrelevant. The physician's argument has stopped your heart.

(f) *A Medical-Test Paradox*

A certain disease is present in 1 out of 500 people in a certain population. A test for the disease exists that is 100% accurate for those who actually have the disease. The test gives a false positive result for 5% of people who do not have the disease. A person is selected at random from the population and given the test. If the test result is positive, what is the probability that the person actually has the disease?

SOLUTION. Let D, $\overline{D}$, and T denote the events:

D = "person has the disease"

$\overline{D}$ = "person does not have the disease"

and

T = "test result is positive."

Notice that $\overline{D}$ is the complement of D so that $D \cap \overline{D} = \emptyset$.

We are given that $P(D) = 1/500 = .002$. Thus $P(\overline{D}) = .998$.

We are also given the conditional probabilities $P(T \mid D) = 1$ and $P(T \mid \overline{D}) = .05$.

We want to find $P(D \mid T)$.

Clearly $T = (T \cap D) \cup (T \cap \overline{D})$. Since D and $\overline{D}$ are disjoint so are $T \cap D$ and $T \cap \overline{D}$. Thus $P(T) = P(T \cap D) + P(T \cap \overline{D})$. The properties of conditional probability from chapter 7 then give:

$$
\begin{aligned}
P(D \mid T) &= \frac{P(D \cap T)}{P(T)} \\[2mm]
&= \frac{P(D \cap T)}{P(T \cap D) + P(T \cap \overline{D})} \\[2mm]
&= \frac{P(D \cap T)}{P(D \cap T) + P(\overline{D} \cap T)} \\[2mm]
&= \frac{P(D)P(T \mid D)}{P(D)P(T \mid D) + P(\overline{D})P(T \mid \overline{D})} \\[2mm]
&= \frac{(.002)(1)}{(.002)(1) + (.998)(.05)} \\[2mm]
&= .0389.
\end{aligned}
$$

Thus, even if you test positive for this disease, the probability you have it is quite small.

Chapter 10

SUMMING UP

Long ago my professor told me what I was like. We talked in his campus office. He sat leaning back in his chair, feet up easy on his desk. I stood near the chalkboard, not quite at military attention. We had talked pure mathematics for an hour and then we discussed the progress, or lack of progress, of my doctoral dissertation. Now, as always, he was ending our discussion with a different kind of instruction.

Our weekly meetings consisted mostly of informal lectures delivered from me to him. I would write my new mathematics on the chalkboard. He would listen and criticize and offer suggestions. These sessions were intense and filled with apprehension and I measured the distance to my doctorate precisely by his reaction to the mathematics I put before him.

But, when my talk ended, he would invariably close the show with a lecture of his own, this one nontechnical and from him to me. The topics varied; you could not predict. My professor was a fine mathematician and from him I learned how to do mathematics. But, like many research mathematicians, his nontechnical interests were somewhat eccentric. Once he told me about Chinese agricultural methods for soybean growing. And once he described certain bizarre psychological characteristics he believed common to all academic administrators. Another time he explained in great detail his secret technique for swimming long distances underwater. But on this particular day he was being philosophical. This day, we had a rare dialogue.

"Mathematics graduate students," he said, "are the brightest and the laziest students in any university."

"Really?" I said, half pleased.

"Of course, there are exceptions to the 'bright' part," he said looking me straight in the eye.

I was no longer even half pleased. "Why do you say mathematics graduate students are lazy?"

"Because they have discovered that to learn mathematics they need only learn a few fundamental things. The rest of mathematics comes from these few things."

"Why does that make them lazy?"

"Lazy by comparison," he said. "Look at other fields. A graduate student of literature cannot even begin work until he has read himself all the way around the walls of a library. And a chemist toils for months in a laboratory before his experiments even run properly, let alone produce anything new. Mathematics students do none of this."

"What must a mathematics student do?" I asked.

"Understand," my professor said.

"OK," I said, conceding. "Mathematics graduate students are lazy. But how do you know they are bright?"

"Because they've learned the secret."

"Secret?"

"Yes," he said. "They've learned that mathematics is pretty."

"How can you be sure they've learned it?"

"Because they are here," he said. "There is no other reason to study mathematics at this level."

"Why does that make them bright?"

"Because," he said, "they learned the secret by themselves. They took dozens of mathematics courses and they were told over and over that mathematics is useful. But no one told them it is beautiful. They learned that by themselves, one by one, all alone."

"But mathematics is useful."

"Certainly," he said. "Without mathematics there would be no technology, no modern world. But that's not why you do it. You do mathematics because it's pretty."

"And that's the secret they've learned?"

"Yes," he said. "They learned it alone and that makes them bright."

"How about me?" I asked. "I learned it too."

"Yes, but not by yourself."

"Who told me?"

"I did," he said, smiling. "Just now."

I did not then know what to make of the "bright but lazy" stuff my professor told me that day in his office. Nor do I now, all these years later. But I recognized even then that he was onto something no matter how casually he laid it out. And looking back I see that he came close to setting down for me the exact characteristics that separate mathematics from the other academic subjects and divide mathematicians from the rest of intellectual society.

But he did not set it down and years passed before I understood. I now think I do. I wish I could tell him I came to it myself the way his bright graduate students came alone to the secret of mathematics. But I'm not sure I did. Just as Moby Dick became part of each sea he swam, I am now touched by all I've done and by many of those I've met. My flanks are as scarred and barnacled as the white whale's. I cannot always be certain of what is my own and what I've learned from others. But—from whatever source—I have over the passing years learned something about mathematics and about mathematicians.

My professor made three points in our long-ago dialogue:

(1) *Beauty motivates mathematics*
(2) *Mathematics flows from basic principles*
(3) *Mathematics possesses utility.*

I made these three points at the beginning of our study. There, I asserted that points (1) and (2) characterize mathematicians in the sense that, among all the members of the intellectual world, only the mathematicians understand the true meaning of these two facts. Mathematicians also understand and appreciate point (3), as do many others. But it is not utility that motivates the creation of mathematics. My professor said it truly and well:

You do mathematics because it is pretty.

A phrase used by the contemporary mathematician Sherman Stein wonderfully condenses points (1), (2), and (3). Stein speaks of:

an appreciation of the beauty, extent, and vitality of mathematics.[1]

Stein's *beauty* meshes exactly with point (1) above. I will interpret his *vitality* as "capacity to grow or develop." And I take *extent* to mean "the range over which something extends."

Thus the *vitality of mathematics* refers to the subject's ability to develop from a handful of embryonic principles into something as colossal and over-reaching as the great mathematical world itself. And the phrase the *extent of mathematics* suggests the enormity of this reach as it becomes the very language of Galileo's book of nature.

I wrote *The Art of Mathematics* to expose the concept of mathematical beauty. My specific purpose was to demonstrate the validity of the assertion: *beauty motivates mathematics*. I wanted to bring my professor's secret out into the sunlight. The present book considers mainly the *vitality* of mathematics. We began with only a few fundamental notions having to do with sets and counting. From these we moved, in an essentially self-contained manner, to some of the basic concepts of number theory, algebra, probability theory, and calculus. We saw mathematics grow from embryo to something reasonably adult.

But the concept of *extent* was always close at hand. We always kept before us the picture of the real world and the mathematical world side by side and the passing of mathematics from one to the other. Somehow the mathematical objects manage to ride an imaginary ether from their airy domain back into the real world where they predict and explain the behavior of concrete phenomena.

Perhaps this omnipresent picture yields an appropriate metaphor for the extent or the reach of mathematics. Give the word *reach* its old nautical meaning and then *mathematics becomes an abstract wind abeam with the world*.

THE AESTHETIC PRINCIPLE OF UTILITY

Physicist John Polkinghorne describes the essence of mathematics when he writes:

> Mathematics is the abstract key that turns the lock of the physical universe.[2]

Here, in a single sentence, Polkinghorne sets down the quintessential nature of mathematics and of the universe, and he describes the relation between the two. Properly read, Polkinhorne's sentence says:

(1) *Mathematics is abstract.*
(2) *The universe is not abstract.*
(3) *Mathematics explains and predicts physical phenomena.*

No one who has thought deeply about this doubts Polkinghorne's assertion. The long march of mathematicians and scientists from Archimedes' time until today has concurred. Science exists for the purpose of the formulation of theories that explain the past, help us understand the present, and in some small sense, predict the future. Scientists formulate their theories by a process that combines observation and extrapolation. They listen to the cosmos. And then they write their theories in mathematics because mathematics is the language that the cosmos speaks.

The process is always the same. A scientist paints, in the mathematical world, an abstract picture of real-world phenomena. This gives him a mathematical model of a portion of the real world. Next he does pure mathematics on the model. Then he extrapolates the new truths, yielded up by the mathematics, back to the real world. Polkinghorne says the process works. So do many others.

The physicist E. P. Wigner, you will remember, believes the effectiveness of mathematics is unreasonable. Martin Gardner, as we have seen, considers it only reasonable. Despite Gardner's brillance and eminence, I concur with Wigner. Another who seems to concur is the physicist Freeman Dyson. In his book *Disturbing the Universe*, Dyson describes the pleasure he found in, for the first time, being allowed to do some serious laboratory experiments. Until this time, his work had been theoretical and mathematical. He watched an electron move on an oil drop, saw it move exactly as mathematics said it should move. He watched in wonder:

> Why it is so, why the electron pays attention to our mathematics, is a mystery even Einstein could not fathom.

Never mind. Unreasonable or not, mathematics works.

Of course not every one thinks so. An old college pal had in his youth an interest in games of chance. He asked me one day about coin tosses.

"What is the probability a coin will land heads 100 times in a row?"

"If the coin is fair," I said, "the probability is $\left(\dfrac{1}{2}\right)^{100}$."

"That's pretty small, isn't it?"

"Tiny," I said. "$\left(\dfrac{1}{2}\right)^{100}$ is virtually zero. In decimal form there will be about 30 zeros following the decimal point."

"Then it's not likely to get 100 straight heads."

"Not likely," I said.

"OK," he said. "But suppose this unlikely event has happened. Suppose the coin has landed heads 100 straight times. What's the probability it comes up heads on the next toss?"

"One half," I said.

"What?" he said.

"One half," I repeated. "I'm assuming the coin tosses are independent of each other. If so, then the probability the coin lands heads on the 101st toss, given that it has landed heads on the first 100 tosses, is exactly ½."

"That's crazy," he said. "The coin surely remembers the many times it has landed heads. When the string gets long enough, it switches to tails."

"How do you know when this will happen?" I asked.

"A real gambler knows."

"The art of gambling, I suppose."

"Exactly. But you should never gamble," he told me. "You'll never make money at it."

And I never did. But, then, neither did he.

In those days my pal knew little mathematics and I was not surprised at his skepticism. Among the cognoscente, however, there is no skepticism. Mathematics turns the lock.

What do you do, however, if mathematics works too well? Suppose, for example, two scientists want to study a certain real-world phenomenon. After a certain amount of observation and reflection, each scientist follows the applied mathematics process described in figure 16 of chapter 1. Scientist A builds a mathematical model and, independently, so does scientist B.

Let's denote the phenomenon under consideration by the symbol P. Let M_1 and M_2 stand, respectively, for the mathematical models of scientist A and scientist B.

Each scientist works through the analysis stage of figure 16. Thus each scientist produces a collection of mathematical calculations that reveals truths about the respective models that were not known at the outset. When this is done, each scientist goes back to the real world and, through experimentation and observation, determines whether or not the new truths agree with real-world observation. That is, each scientist determines whether or not the theoretical model agrees with real-world observation.

We have here a potentially perplexing situation. It is possible that both models agree with observation and yet the models themselves differ from one another, and perhaps differ radically. Moreover, it may be that model M_1 and model M_2 offer vastly different predictions about future behavior, or about nonobservable behavior, of phenomena P. Which model then do we choose? Which is correct?

An important example of this situation occurred not all that long ago. In my graduate student days there existed two highly regarded and competing cosmological theories. The first of these—the big bang theory—remains before us. In this theory, as you know, the universe begins with a single explosion, a great fireball known as the big bang. In this theory the entire cosmos is born all at once, at a single instant. The second theory—called the steady-state theory—suggested a radically different scenario. In the steady-state theory, the creation of matter occurs continuously. Rather than the explosive, point-singularity birth suggested by the big bang theory, steady-state proposed a universe in which matter is being constantly created. Select a liter of universe out there somewhere in the void. Watch it carefully. Sooner or later a hydrogen atom will pop into existence.

Obviously, the two theories are inconsistent. They cannot both hold. Yet for a considerable period, scientists were unable to determine with finality which was correct. Both theories—in their observable aspects—agreed with what the scientists could see and measure. Thus for many years we had under consideration the exact situation described above: two vastly differing models M_1 and M_2 purporting to explain nothing less than the creation of universal matter. And we could not by observation choose between them. Only around 1965—when the discovery of background cosmic radiation provided a faint echo of the great blast—was the steady-state theory discredited.

So, in the case of steady-state versus big bang, observational evidence finally appeared that allowed scientists to choose one model over the other. But what if no such evidence had come forth? How do you decide between M_1 and M_2 in the absence of discriminating observational evidence? It's easy. You choose the more beautiful of the two.

The race may not always go to the swift and the battle to the strong, but that's how you bet. And the more beautiful of the two models may not be correct, but that's how you choose. Others will tell you the same thing. Listen to two famous physicists, Paul Dirac and Werner Heisenberg. Dirac said:

> It is more important to have beauty in one's equations than to have them fit experiment . . . [3]

And Heisenberg:

> If nature leads us to mathematical forms of great beauty and simplicity . . . we cannot help thinking that they are true, that they reveal a genuine feature of nature.[4]

These two scientists tell you to choose the loveliest model whenever you can. Now, the models themselves are made of mathematics. And the mathematics itself, as we have seen, is motivated by aesthetics. The highest praise one mathematician can give to another is to pronounce the other's work *elegant*. G. H. Hardy put it this way:

> Beauty is the first test: there is no permanent place in the world for ugly mathematics.[5]

So, we have a situation where mathematicians create mathematics for aesthetic reasons and then essentially allow the nonlovely mathematics to slide away. Then the scientists build their models of reality using the resulting mathematics. And, in addition, they choose between competing models—when they must—with aesthetic judgments. Thus the concept of beauty plays both a primary and a secondary role in the application of mathematics to real-world problems: first in the mathematics itself and then in the evaluation of competing models of reality. Put all this together and a startling principle falls out.

The aesthetic principle of utility: The most beautiful theo-
ries are the most useful.

As for steady-state versus big bang, the above principle gives the winner ahead of observation. It follows from the mathematics of the steady-state theory that, if matter is constantly created, then the rate of creation can be computed. This rate turns out to be a certain positive real number. Let's denote this number by r. But why is the rate this particular value? Why r? Why should the universe settle on this particular continuous rate of creation of matter? Perhaps it did. But if I have a choice, I'd prefer matter to be created all at once. It seems to me much prettier. And the existence of the big bang theory gives me a choice. Thus big bang is preferable to steady-state for aesthetic reasons.

Of course, there is another possibility. Steady-state gives us rate of creation r. Big bang gives us matter all at once, an infinite rate of creation. How about $r = 0$? This seems to me even lovelier. I prefer the following cosmological principle.

The principle of the static universe: The universe is infi-
nitely old. Matter was never created; it has always existed.

Any physicist will tell you immediately that the static universe is not a tenable possibility. It fails in many ways to agree with observation. No doubt the physicists are correct. Still there is about the static principle a certain minimal elegance.

MIRROR IMAGES

The extent of mathematics then—like the motivation for mathematics—is deeply intertwined with aesthetics. The aesthetic principle says "beauty implies utility." Thus Polkinghorne's abstract key is shaped by the mathematician's concept of beauty. Poets similarly shape their poems. A great poet, John Keats, put it this way:

"Beauty is truth, truth beauty"—that is all
Ye know on earth, and all ye need to know.

Academic administrators, in particular, and much of humankind, in general, tend to think of mathematics as some kind of science. Mathematics departments are traditionally housed within universities as part of a college, or at least a division, of science. This division ordinarily contains the traditional departments of physics, chemistry, biology, and mathematics. And if you go outside the university and stop one hundred random citizens and ask: "Is mathematics an art or is it a science?" you are likely to receive one hundred votes for science. Yet it is almost obvious that mathematics cannot be science. There are several reasons.

(1) *Science is inductive; mathematics is deductive.*
(Scientists conclude, from the observational consistency of past events, the likelihood of future occurrences. Mathematicians deduce mathematical truth by logical argument from known mathematical facts.)

(2) *A theory is a scientific theory only if it is possible to show it to be false.*
(In late afternoon with my office door closed and the blinds drawn so that no one can see or hear, I can wiggle my finger and a copy of *The Complete Shakespeare* floats off my bookcase, and lands softly on my desk, opened to act 1, scene 1 of *Macbeth*. This is a nice theory of motion—a book moves when I point my finger—but it is not a scientific theory since you cannot show it is to be false. It works only when nobody is looking.)

(3) *A theory is a mathematical theory only if it has been proven true.*
(A proven mathematical theory has a name; it is called a theorem.)

Mathematics—we know—is necessary for science. But mathematics is not science. These two subjects are like the two men on the stairway in the famous M. C. Escher print. They walk in the same direction with their feet on the same step. But one goes up, the other down. "Contact between them is impossible," says Escher, "for they live in different worlds."

It is possible that a particular area of science may itself be an art just as mathematics is art. Mathematics, however, is not science.

The art nearest mathematics is the art of poetry. We know they both deal in beauty. But they have other common characteristics. For example:

(a) *Poetry is scary; mathematics is scary.*
(Many veteran actors avoid Shakespeare because speaking in verse frightens them. And everyone knows of the ubiquitous anxiety caused by fear of mathematics.)

(b) *Poetry is difficult; mathematics is difficult.*
(Examples abound. Here are two—one from each subject:

(i) Shakespeare writes:

> Sith every action that hath gone before,
> Whereof we have record, trial did draw
> Bias and thwart, not answering the aim
> And that unbodied figure of the thought
> That gave't surmised shape.

As it stands this is difficult to understand. And, even if I tell you that the lines are from *Troilus and Cressida* and the speaker is Agamemnon, the difficulty remains. *Poetry is difficult.*

(ii) Look again at the epsilon-delta definition of the notion of limit given back in chapter 8. Write the definition carefully on a 3-inch-by-5-inch card. Carry the card next to your heart. Read it thoughtfully several times each day. When you truly understand it, try your hand at writing its negative. That is, write a precise statement of *what it means for the definition to be false.* Then see how long it takes you to understand the new statement. It took me about a year. *Mathematics is difficult.*)

(c) *Poetry is not natural; mathematics is not natural.*
(Both poetry and mathematics use language that is heightened, rich, and condensed. People these days do not naturally speak or write in verse. Nor do they naturally speak or write mathematics. Both poetry and mathematics must be *learned* and a taste for them must be *developed.* It is a fallacy to believe that either poetry or mathematics can be made naturalistic.)

A mathematician sees beauty in abstraction. But he also sees reality. A differential equation is a mathematical object and thus is completely abstract, like all such objects. And there are plenty of them, at least an uncountable infinity. But, by selecting a particular equation from this enormous set, a mathematician can build a model and thus describe the motion of a falling raindrop. In the abstract equation, the mathematician sees the reality of the raindrop's motion.

A poet works the other way around. When Robert Frost wanted to describe a spiritual experience, he wrote about stopping his horse on a snowy evening in a dark wood. Shakespeare placed infinite space inside a nutshell—or would have, had Hamlet not had bad dreams. Similarly Blake, the most antirational of poets, wrote:

> To see a World in a Grain of Sand
> And a Heaven in a Wild Flower,
> Hold Infinity in the palm of your hand
> And eternity in an hour.[6]

In the poet's creative hands the abstract notions of religion, infinite space, heaven, and eternity boil down to considerations of dark woods, nutshells, wild flowers, and an hour of time.

Both the mathematician and the poet trade in beauty and in abstraction. But they come at them in reverse order.

> *A mathematician sees all concrete images embodied in abstract ideas,*

while

> *a poet sees all abstract ideas embodied in concrete images.*

What then is the difference between the mathematician and the poet? Fundamentally, not much. They are face-to-face likenesses of one another, mirror images. Mathematicians know this instinctively. A mathematician looks in a mirror and a poet looks back. In a newer world it will also work the other way around.

NOTES

INTRODUCTION

1. Jerry P. King, *The Art of Mathematics* (New York: Plenum, 1992).
2. C. P. Snow, *The Two Cultures and a Second Look* (London: Cambridge University Press, 1976), p. 4.
3. King, *The Art of Mathematics*, p. 294.

CHAPTER 1

1. William Shakespeare, *A Midsummer Night's Dream*, act 5, scene 1.
2. *Mathematical Reviews*, published by the American Mathematical Society, contains brief reviews of mathematics research papers. As of November 2007, this journal contained information on more than 2.2 million articles.
3. Martin Gardner, "Making No Apologies," *Nature* 358 (July 1992).
4. Alfred Renyi, *Dialogues on Mathematics* (San Francisco: Holden-Day, 1967), p. 11.
5. Robert Frost, "The Secret Sits," in *The Poetry of Robert Frost* (New York: Holt, Rinehart and Winston, 1969), p. 362.
6. Bertrand Russell, *Mysticism and Logic* (Garden City, NY: Doubleday-Anchor), p. 71.
7. E. P. Wigner, "The Unreasonable Effectiveness of Mathematics in the Natural Sciences," *Comunications on Pure and Applied Mathematics* 13 (1980): 1–14.

CHAPTER 2

1. George Gamow, *One, Two, Three, . . . , Infinity* (New York: Viking Press, 1947).
2. Bertrand Russell, *Mysticism and Logic* (Garden City, NY: Doubleday-Anchor), p. 57.
3. Leopold Kronecker, "Uber den Zahibegriff," *Jour. für Math.* 101 (1887): 337–55.

4. Georg Cantor, "Uber eine Eigenschaft der Inbegriffes aller reellen algebraischen Zahlen, *Jour. für Math.* 77 (1874): 258–62.

CHAPTER 3

1. John Milton, "Paradise Lost," VIII, in Milton's *Complete Poems* (New York: Appleton-Century-Crofts, 1933), p. 279.
2. Edmund Landau, *Foundations of Analysis* (New York: Chelsea, 1951).
3. William Shakespeare, *The Winter's Tale*, act 1, scene 2.

CHAPTER 4

1. George Simmons, *Calculus with Analytic Geometry* (New York: McGraw-Hill, 1985), p. 838.
2. Robert Frost, "Birches," in *The Poetry of Robert Frost* (New York: Holt, Rinehart and Winston, 1969), p. 121.
3. Archibald MacLeish, "The End of the World," in *The New Pocket Anthology of American Verse* (New York: Pocket Books, 1955), p. 296.
4. William Shakespeare, *Macbeth*, act 1, scene 2.

CHAPTER 5

1. George Simmons, *Calculus with Analytic Geometry* (New York: McGraw-Hill, 1985), p. 765.
2. Will Durant, *The Story of Civilization*, part II, "The Life of Greece" (New York: Simon & Schuster, 1939), p. 166.
3. Bertrand Russell, *Wisdom of the West* (Garden City, NY: Doubleday, 1959), p. 38.
4. G. H. Hardy, *A Mathematician's Apology* (London: Cambridge University Press, 1973), p. 92.
5. Edmund Landau, *Foundations of Analysis* (New York: Chelsea, 1951).
6. King James Bible, 1 Kings 7:23.
7. Alfred, Lord Tennyson, *Ulysses' Six Centuries of Great Poetry* (New York: Dell, 1955).
8. Herb Silverman, *Complex Variables* (Boston: Houghton Muffin, 1975), p. 72.

CHAPTER 6

1. E. T. Bell, *Men of Mathematics* (New York: Simon & Schuster, 1965), p. 54.
2. Ibid., p. 55.

CHAPTER 7

1. Alexander Pope, *Poetry and Prose of Alexander Pope* (Boston: Houghton Mifflin, 1969), p. 274.
2. Northrop Frye, *Fearful Symmetry* (Boston: Beacon, 1958), p. 50.

CHAPTER 8

1. Salomon Bochner, *The Role of Mathematics in the Rise of Science* (Princeton, NJ: Princeton University Press, 1969), p. 309.
2. Ibid., p. 167.
3. Morris Kline, *Mathematics in Western Culture* (New York: Oxford University Press, 1953), p. 229.
4. Ibid., p. 232.
5. Phillip E. B. Jourdain, "The Nature of Mathematics," in *The World of Mathematics* (New York: Simon & Schuster, 1956), p. 56.
6. Gustav Fechner, *Vorschule der Aesthetik* (Leipzig: Breitkoph and Hartel).
7. I. C. McManus, D. Edmonson, and J. Rodger, "Balance in Pictures," *British Journal of Psychology* 71 (1980): 502–24.

CHAPTER 9

1. G. H. Hardy, *A Mathematician's Apology* (London: Cambridge University Press, 1973), p. 4.
2. Ibid., p. 81.
3. Ernest Hemingway, *Green Hills of Africa* (New York: Charles Scribner's Sons, 1963), p. 109.
4. John Tierney, "Behind Monte Hall's Doors, Debate and Answer," *New York Times*, July 21, 1991, p. 8.

5. Martin Gardner, *The Second Scientific Book of Mathematical Puzzles and Diversions* (New York: Simon & Schuster, 1961), p. 226.

6. Cohn R. Blyth, "Some Probability Paradoxes in Choice from among Random Alternatives," *Journal of the American Statistical Association* (June 1972): 366–73.

7. Arthur Conan Doyle, "The Adventure of the Speckled Band," in *The Complete Sherlock Holmes*, vol. 1 (New York: Doubleday, 1930), p. 272.

CHAPTER 10

1. Sherman K. Stein, *Mathematics: The Man-Made Universe* (San Francisco: Freeman, 1976), p. xiv.

2. John Polkinghorne, *One World: The Interaction of Science and Theology* (Princeton, NJ: Princeton University Press, 1987), p. 46.

3. Paul Dirac, "The Evolution of the Physicists's Picture of Nature," *Scientific American* (May 1963).

4. Werner Heisenberg, *Physics and Beyond* (New York: Harper & Row, 1971), p. 68.

5. G. H. Hardy, *A Mathematician's Apology* (London: Cambridge University Press, 1973), p. 85.

6. Blake, "Auguries of Innocence," in *The Laurel Poetry Series: Blake* (New York: Dell, 1960), p. 99.

KEY TO SYMBOLS

Symbol	How to read it	Defined on page	Intuitive description
$p \Rightarrow q$	p implies q.	31	Statement q follows from statement p.
$p \Leftrightarrow q$	p if and only if q.	40	p and q are equivalent statements.
$p \wedge q$	p and q.	34	True only when both p and q are true.
$p \vee q$	p or q.	35	True when one or both of p and q are true.
$\sim p$	Not p.	36	True only when p is false.
dv/dt	Derivative of v with respect to t.	309	Rate of change of v with with respect to t.
S	The set S.	56	S is a collection of objects.
$x \in S$	x belongs to S	57	x is a member of the set of objects named S.
$x \notin S$	x does not belong to S.	57	x is not a member of the set of objects named S.
$\{x : P(x)\}$	Set of all x such that $P(x)$ is true.	58	The set of all x's for which statement $P(x)$ is true.
$\mathbb{N}$	The set of natural numbers $\{1, 2, 3, \ldots\}$.	59	The set that contains the numbers 1, 2, 3, 4, and so on.
$\varnothing$	The empty set.	59	The set that contains no members.
$T \subset S$	T is a subset of S.	59	Each member of T also belongs to S.

$n(S)$	The number of elements in S.	64	The total number of members of S.
$\mathbb{R}$	The real numbers.	195	The set that contains all the rational numbers and all the irrational numbers.
$S \cup T$	The union of S and T.	70	The set of all objects that belong to S or T or both.
$S \cap T$	The intersection of S and T.	71	The set of all objects that belong to both S and T.
$S \setminus T$	The complement of T in S.	72	The set of all objects that belong to S but do not belong to T.
$\sim T$	The complement of T.	73	The set of all objects under consideration that do not belong to T.
$n < m$	n is less than m.	93	The number n is strictly smaller than the number m.
$n \leq m$	n is less than or equal to m.	94	The number n either equals the number m or is strictly smaller.
$\mathbb{Z}$	The set of integers.	96	The set that contains the numbers 1, 2, 3, ..., their negatives, $-1, -2, -3, \ldots$, and the number 0.
$\mathbb{Z}^+$	The set of positive integers.	96	The set that contains the numbers 1, 2, 3, ...; just another name for the set of natural numbers.
$\dfrac{n}{m}$	A rational number.	109	A number that is the quotient of two integers.

$\mathbb{Q}$	The set of rational numbers.	115	Set of all numbers that are quotients of integers.
$d \mid n$	d divides n.	139	The integer d divides into the integer n and leaves no remainder.
$\sqrt{n}$	Square root of n.	146	The nonnegative number whose square is n.
$gcd\,(a, b)$	Greatest common divisor of a and b.	155	The largest number that divides both a and b.
$a \equiv b \bmod m$	a is congruent to b modulo m.	160	The difference between a and b is divisible by m.
$a \sim b$	a is related to b.	166	The numbers a and b are related in some given way.
$C(x)$	The equivalent class determined by x.	166	The set of numbers that are each related to x.
$\bar{n}$	The set of integers congruent to n.	175	Another symbol for the set of numbers that are related to n.
$a \star b$	A binary operation on a and b.	172	An operation that turns two members of a set into a third member.
$\mathbb{C}$	The set of complex numbers.	212	The set of all numbers $\mathbb{Z}$ of the form $z = x + iy$.
i	The complex number whose square is negative one.	213	The complex number whose square is -1.
$f : A \to B$	f is a function from A into B.	219	A rule that associates, with each member of A, a unique member of B.

$f(x)$	f of x.	219	The value of the function f at x.
$\mathcal{D}(f)$	The domain of f.	219	The set of all objects for which f is defined.
$\mathcal{R}(f)$	The range of f.	220	The set of all values of the function f.
$\mathcal{G}(f)$	The graph of f.	226	Curve or line that describes the behavior of the function f.
$d(P_1, P_2)$	The distance between P_1 and P_2.	232	Length of the straight line joining the two given points.
$f \circ g$	The composition of f and g.	235	The function obtained by first using g and then f.
f^{-1}	The inverse of f.	236	The function obtained by reverse use of f.
$P(A)$	The probability of A.	244	Probability that the event A occurs.
$w(x)$	The weight of x.	245	Weight assigned to a member of a sample space.
$\displaystyle\sum_{k=1}^{m} t_k$	The sum from $k = 1$ to $k = m$ of t sub k.	246	Result of adding the indicated numbers.
$\displaystyle\sum_{x \in B} f(x)$	The sum of $f(x)$ over all x in B.	247	Result of adding all the values of $f(x)$ for x in B.
$P(A|B)$	Conditional probability of A given B.	259	The probability that A occurs when B has already occurred.
$n!$	n factorial.	266	The product of all the natural numbers between 1 and n.

$P(n, r)$	Number of permutations of n objects taken r at a time.	267	The number of ways to arrange n objects on a line when they are taken r at a time.
$C(n, r)$	The number of subsets of size r that can be chosen from a set of size n.	269	The number of combinations that can be made of n objects when they are taken r at a time.
$\binom{n}{r}$	Binomial coefficient.	270	Another symbol for $C(n, r)$.
$E(f)$	Expected value of f.	282	The expected value of a probabilistic experiment.
$\lim_{n \to \infty} A_n$	The limit of A_n as n tends to infinity.	296	The value that A_n approaches as n increases forever.
$\int_a^b f(x)dx$	The integral from a to b of f of x, dee x.	303	The area under the graph of f between a and b.
$f'(x)$	The derivative of f with respect to x.	310	The slope of the line tangent to the graph of f at x.
$s''(t)$	Second derivative of s with respect to t.	309	The acceleration of a particle moving on a straight line.
d^2s/dt^2	Second derivative of s with respect to t.	309	Another symbol for $s''(t)$.
$\lim_{x \to a} g(x) = L$	The limit of $g(x)$ as x approaches a equals L.	324	$g(x)$ is arbitrarily close to L whenever x is sufficiently close to a.

INDEX